RFID Technology and Applications

Are you an engineer or manager working on the development and implementation of RFID technology? If so, this book is for you.

Covering both passive and active RFID, the challenges to RFID implementation are addressed using specific industry research examples as well as common integration issues. Key topics such as performance optimization and evaluation, sensors, network simulation, RFID in the retail supply chain, and testing are covered, as are applications in product lifecycle management in the automotive and aerospace sectors, in anti-counterfeiting, and in health care.

This book brings together insights from the world's leading research laboratories in the field, including MIT, which developed the Electronic Product Code (EPC) scheme that is set to become the global standard for object-identification.

This authoritative survey of core engineering issues, including trends and key business questions in RFID research and practical implementations, is ideal for researchers and practitioners in electrical engineering, especially those working on the theory and practice of applying RFID technology in manufacturing and supply chains, as well as engineers and managers working on the implementation of RFID.

Stephen B. Miles is an RFID evangelist and Research Engineer for the Auto-ID Lab at MIT. He has over 15 years of experience in computer network integration and services.

Sanjay E. Sarma is currently an Associate Professor at MIT, and is also a co-founder of the Auto-ID Center there. He serves on the board of EPCglobal, the wordwide standards body he helped to start up.

John R. Williams is Director of the Auto-ID Lab at MIT, and is also a Professor of Information Engineering in Civil and Environmental Engineering. As well as many years of lecturing, has also worked in industry, and was the Vice President of Engineering at two software start-up companies.

The Auto-ID Lab at MIT has developed a suite of RFID and software specifications for an Electronic Product Code (EPC) network that have been incorporated into EPCglobal and ISO standards and are being used by over 1,000 companies across the globe.

RFID Technology and Applications

Edited by

STEPHEN B. MILES
SANJAY E. SARMA
JOHN R. WILLIAMS
Massachusetts Institute of Technology

CAMBRIDGE UNIVERSITY PRESS

Cambridge, New York, Melbourne, Madrid, Cape Town, Singapore,
São Paulo, Delhi, Dubai, Tokyo, Mexico City

Cambridge University Press
The Edinburgh Building, Cambridge CB2 8RU, UK

Published in the United States of America by Cambridge University Press, New York

www.cambridge.org
Information on this title: www.cambridge.org/9780521169615

First published 2008
First paperback edition 2010

A catalogue record for this publication is available from the British Library

Library of Congress Cataloguing in Publication data

RFID technology and applications / edited by Stephen B. Miles, Sanjay E. Sarma, John R. Williams.
      p.   cm.
   Includes bibliographical references and index.
   ISBN 978-0-521-88093-0 (hardback)
   1. Radio frequency identification systems.   2. Inventory control–Automation.   I. Title.
   TS160.R437 2007
   658.7′87–dc22   2007049093

ISBN 978-0-521-88093-0 Hardback
ISBN 978-0-521-16961-5 Paperback

Contents

List of contributors   *page* xi
Preface   xv
Acknowledgments   xxi

1   Introduction to RFID history and markets   1
      Stephen Miles

      1.1   Market assessment   3
      1.2   Historical background   4
      1.3   Adoption of the Auto-ID system for the Electronic
            Product Code (EPC)   6
      1.4   EPC information services   8
      1.5   Methodology – closing the loop   9
      1.6   RFID investing in a better future   10
      1.7   New business processes   12
      1.8   References   13

2   RFID technology and its applications   16
      Sanjay Sarma

      2.1   The first wave: the state of EPC technology   16
      2.2   On the future of RFID technology   21
      2.3   Applications   25
      2.4   Conclusions   30
      2.5   References   30

3   RFID tag performance optimization: a chip perspective   33
      Hao Min

      3.1   Metrics of tag performance   33
      3.2   Performance enhancement of RFID tags   36
      3.3   Sensors for RFID; integrating temperature sensors
            into RFID tags   44
      3.4   References   46

4   Resolution and integration of HF and UHF   47
Marlin H. Mickle, Leonid Mats, and Peter J. Hawrylak

   4.1   Introduction   48
   4.2   Basics of the technologies   48
   4.3   Fundamentals of orientation   50
   4.4   Antennas and materials   53
   4.5   An analogy to network layering   55
   4.6   Examples of converging technologies   57
   4.7   Technical summary   57
   4.8   Pharma – a surrogate for the future   59
   4.9   References   60

5   Integrating sensors and actuators into RFID tags   61
J. T. Cain and Kang Lee

   5.1   Introduction   61
   5.2   RFID systems   61
   5.3   "Smart" transducers   63
   5.4   RFID tags with sensors   68
   5.5   Conclusion   72
   5.6   Acknowledgment   72
   5.7   References   72

6   Performance evaluation of WiFi RFID localization technologies   74
Mohammad Heidari and Kaveh Pahlavan

   6.1   Introduction   75
   6.2   Fundamentals of RFID localization   76
   6.3   Performance evaluation   80
   6.4   Summary and conclusions   84
   6.5   Acknowledgments   84
   6.6   References   86

7   Modeling supply chain network traffic   87
John R. Williams, Abel Sanchez, Paul Hofmann, Tao Lin, Michael Lipton,
and Krish Mantripragada

   7.1   Introduction and motivation   87
   7.2   Requirements   88
   7.3   Software architecture   91
   7.4   Implementation   93
   7.5   Simulator performance   96
   7.6   References   97
   7.7   Appendix   97

8   Deployment considerations for active RFID systems   101
         Gisele Bennett and Ralph Herkert

         8.1   Introduction   101
         8.2   Basics of the technologies   102
         8.3   Technology and architectural considerations   103
         8.5   Testing for RFID performance and interference   109
         8.6   References   111

9   RFID in the retail supply chain: issues and opportunities   113
         Bill C. Hardgrave and Robert Miller

         9.1   Introduction   113
         9.2   From partial to full supply chain coverage   113
         9.3   Store execution   115
         9.4   Data analytics   118
         9.5   Conclusion   119
         9.6   References   119

**10   Reducing barriers to ID system adoption in the aerospace
         industry: the aerospace ID technologies program   121**
         Duncan McFarlane, Alan Thorne, Mark Harrison, and Victor Prodonoff Jr.

         10.1   Introduction   121
         10.2   Background   121
         10.3   The Aero ID consortium   123
         10.4   Defining a research program   125
         10.5   Research developments   127
         10.6   Trials and industrial adoption   137
         10.7   Summary   142
         10.8   Bibliography   143

11   The cold chain   144
         J. P. Emond

         11.1   The food industry   144
         11.2   Pharmaceuticals   146
         11.3   Types of temperature-tracking technologies   147
         11.4   Challenges associated with RFID temperature-tracking
                technologies   149
         11.5   Potential applications in "semi- and real-time" cold chain
                management   153
         11.6   References   155

12   **The application of RFID as anti-counterfeiting technique: issues**
           and opportunities   157
           Thorsten Staake, Florian Michahelles, and Elgar Fleisch

           12.1   Counterfeit trade and implications for affected enterprises   157
           12.2   The use of RFID to avert counterfeit trade   159
           12.3   Principal solution concepts based on RFID   162
           12.4   Migration paths and application scenarios   166
           12.5   Conclusion   167
           12.6   References   167

13   **Closing product information loops with product-embedded**
           information devices: RFID technology and applications, models and metrics   169
           Dimitris Kiritsis, Hong-Bae Jun, and Paul Xirouchakis

           13.1   Introduction: closing the product information loop   169
           13.2   The concept of closed-loop PLM   171
           13.3   The state of the art   173
           13.4   System architecture   174
           13.5   A business case of PROMISE on ELV recovery   176
           13.6   Product usage data modeling with UML and RDF   177
           13.7   Conclusion   181
           13.8   Acknowledgments   181
           13.9   References   181

14   **Moving from RFID to autonomous cooperating logistic processes**   183
           Bernd Scholz-Reiter, Dieter Uckelmann, Christian Gorldt, Uwe Hinrichs, and Jan Topi Tervo

           14.1   Introduction to autonomous cooperating logistic processes
                  and handling systems   183
           14.2   Radio frequency – key technology for autonomous logistics   185
           14.3   RFID-aware automated handling systems – the differentiator
                  between intelligent objects and autonomous logistics   192
           14.4   Conclusion   195
           14.5   References   195

15   **Conclusions**   198
           Stephen Miles, Sanjay Sarma, and John Williams

           15.1   Radio frequency gap analyses; Georgia Tech LANDmark
                  Medical Device Test Center   199
           15.2   The RFID Technology Selector Tool; Auto-ID Labs at
                  Cambridge University   199
           15.3   An EPC GenII-certified test laboratory; the RFID Research
                  Center, University of Arkansas   200

15.4   ISO 18000-7 and 6c (HF and UHF) RFID and
         EPC network simulation   200
15.5   RFID anti-counterfeiting attack models; Auto-ID Labs at
         St. Gallen and the ETH Zürich   203
15.6   Adding sensors to RFID Systems – IEEE 1451/NIST
         interface specifications   204
15.7   Adding location interfaces   205
15.8   Convergence of RFID infrastructure: multi-frequency and
         multi-protocol   207
15.9   New business processes: from e-Pedigree to VAT tax compliance   208
15.10   References   211

Appendix – links to RFID technology and applications resources   213
Editor biographies   215
Index   217

Contributors

Gisele Bennett

Professor and Director, Electro-Optical Systems Laboratory, Georgia Tech Research Institute, Georgia Institute of Technology, Atlanta, GA – Ch. 8

J. T. Cain

Professor, Department of Electrical and Computer Engineering, University of Pittsburgh, Pittsburgh, PA – Ch. 5

J. P. Emond

Associate Professor and Co-Director, IFAS Center for Food Distribution and Retailing, University of Florida, Gainesville, FL – Ch. 11

Elgar Fleisch

Professor and Director, Auto-ID Labs, University of St. Gallen, Institute of Technology Management and Eidgenössische Technische Hochschule Zürich, Zurich, Department of Management, Technology and Economics – Ch. 12

Christian Gorldt

Bremer Institut für Produktion und Logistik GmbH (BIBA), University of Bremen, Bremen – Ch. 14

Bill C. Hardgrave

Associate Professor and Executive Director, RFID Research Center, Sam M. Walton College of Business, University of Arkansas, Fayetteville, AR – Ch. 10

Mark Harrison

Senior Research Associate, Auto-ID Labs, University of Cambridge, Cambridge – Ch. 10

Peter J. Hawrylak

RFID Center of Excellence, University of Pittsburgh, Pittsburgh, PA – Ch. 4

Mohammad Heidari
Center for Wireless Information Network Studies (CWINS), Worcester Polytechnic Institute, Worcester, MA – Ch. 6

Ralph Herkert
Senior Research Engineer, Medical Device Test Center, Georgia Tech Research Institute, Georgia Institute of Technology, Atlanta, GA – Ch. 8

Uwe Hinrichs
Bremer Institut für Produktion und Logistik GmbH (BIBA), University of Bremen, Bremen – Ch. 14

Paul Hofmann
Director External Relations, SAP Research Center, Palo Alto, CA – Ch. 7

Hong-Bae Jun
Ecole Polytechnique Fédérale de Lausanne, Lausanne – Ch. 13

Dimitris Kiritsis
Associate Director, Laboratory of Informatics for Design and Production, Ecole Polytechnique Fédérale de Lausanne, Lausanne – Ch. 13

Kang Lee
National Institute of Standards and Technology (NIST), Washington, DC – Ch. 5

Tao Lin
Director of Auto-ID Infrastructure, SAP Labs LLC, Palo Alto, CA – Ch. 7

Michael Lipton
Director of RFID Solutions Management for Healthcare Life Sciences, SAP Labs LLC, Palo Alto, CA – Ch. 7

Duncan McFarlane
Professor and Director, Auto-ID Lab, University of Cambridge, Cambridge – Ch. 10

Krish Mantripragada
Global Lead and Program Director, Supply Chain Solutions, SAP Labs LLC, Palo Alto, CA – Ch. 7

Leonid Mats
RFID Center of Excellence, University of Pittsburgh, Pittsburgh, PA – Ch. 4

Florian Michahelles
Department of Management, Technology and Economics, Eidgenössische Technische Hochschule Zürich, Zurich – Ch. 12

Marlin H. Mickle
Professor and Director, RFID Center of Excellence, University of Pittsburgh, Pittsburgh, PA – Ch. 4

Stephen Miles
Research Engineer and Co-Chair, the RFID Academic Convocation, Auto-ID Labs, Massachusetts Institute of Technology, Cambridge, MA – Ch. 1, 15

Robert Miller
Assistant Professor, Accounting and MIS Department, Dauch College of Business & Economics, Ashland University, Ashland, OH – Ch. 10

Hao Min
Professor and Director Auto-ID Labs, Fudan University, Shanghai – Ch. 3

Kaveh Pahlavan
Professor and Director, Center for Wireless Information Network Studies (CWINS), Worcester Polytechnic Institute, Worcester, MA – Ch. 6

Victor Prodonoff Jr.
Auto-ID Labs, University of Cambridge, Cambridge – Ch. 10

Abel Sanchez
Research Scientist, Auto-ID Labs, Massachusetts Institute of Technology, Cambridge, MA – Ch. 7

Sanjay Sarma
Associate Professor, Massachusetts Institute of Technology, Auto-ID Labs, Cambridge, MA – Ch. 2

Bernd Scholz-Reiter, Ph.D.
Professor and Director, Bremer Institut für Produktion und Logistik GmbH (BIBA), University of Bremen, Bremen, – Ch. 14

Thorsten Staake
Massachusetts Institute of Technology, Auto-ID Labs, Cambridge, MA – Ch. 12

Jan Topi Tervo
Bremer Institut für Produktion und Logistik GmbH (BIBA), University of Bremen, Bremen – Ch. 14

Alan Thorne
Auto-ID Lab, University of Cambridge, Cambridge – Ch. 10

Dieter Uckelmann
Research Scientist and Manager RFID-Application and Demonstration Center, Bremer Institut für Produktion und Logistik GmbH (BIBA), University of Bremen, Bremen – Ch. 14

John R. Williams
Associate Professor and Director, Auto-ID Labs, Massachusetts Institute of Technology, Cambridge, MA – Ch. 7

Paul Xirouchakis
Ecole Polytechnique Fédérale de Lausanne, Lausanne – Ch. 13

Preface

This book is addressed to business management and project managers as well as researchers who are evaluating the use of radio frequency identification (RFID) for tracking uniquely identified objects. In an effort to make RFID project management less of an art form and more of a science *RFID Technology and Applications* brings together pioneering RFID academic research principals to analyze engineering issues that have hampered the deployment of RFID and to share "best practices" learnings from their work. By extending the original work of the Auto-ID Center at MIT and the subsequent Auto-ID Labs consortium led by MIT that now comprises seven world-renowned research universities on four continents, this book seeks to establish a baseline for what RFID technology works today and identifies areas requiring research on which other researchers in academic, commercial, and regulatory agencies can build.

The researchers represented in these pages have gathered on three continents in the course of the RFID Academic Convocations, a research collaboration hosted by the Auto-ID Labs that started in January of 2006, at MIT, and was followed by events co-hosted with the Chinese Academy of Sciences and Auto-ID Labs at Fudan University in Shanghai, as RFID Live! 2007 pre-conference events, and by the event in Brussels organized with the European Commission Directorate-General for Informatics (DGIT) and the Auto-ID Labs at Cambridge University. These Convocations bring together academic researchers with industry representatives and regulatory stakeholders to collaborate across disciplines and institutions to identify challenges faced by industry in adopting RFID technology. As summarized by Robert Cresanti, Under Secretary of Commerce for Technology, United States Department of Commerce in his remarks that day, "the two primary challenges facing this new technology are standards and interoperability issues across various RFID systems, companies, and countries, and privacy and security concerns."[1]

Following an introduction to the history of RFID as it bears on standards and interoperability, the technology chapters that follow (Chs. 2–7) address core engineering issues related to the design of RFID chips and antennas that must be tuned to specific products, the placement, packaging, and density of those tags to

---

[1] Technology Administration Speech, remarks by Robert C. Cresanti, Under Secretary of Commerce for Technology, United States Department of Commerce, delivered March 13, 2007 at the EU RFID FORUM 2007 (http://www.technology.gov/Speeches/RC_070313.htm).

maximize their readability, and the characterization of downstream RF operating environments, and the reader range and densities for effective (read accuracy and speed) RFID data acquisition and secure information exchange.

In investigating RFID applications (Chs. 9–15) researchers illustrate the challenges of implementing RFID applications today, especially where they are seeking to change current business processes. Sanjay Sarma, co-founder of the Auto-ID Center at MIT and EPCglobal board member, leads the RFID technology section (Chs. 2–7) with an introduction to the technology that he was personally instrumental in developing at the Auto-ID Center and subsequently as interim CTO and Board Member for Oat Systems, a leading RFID middleware company founded by his graduate student Laxmiprasad Putta. Sanjay Sarma sets the stage for the subsequent technology chapters by highlighting the many areas of ongoing research related to RFID (Ch. 2). The introduction to designing RFID tags optimized for low power consumption by Hao Min, Director of the Auto-ID Labs at Fudan University (Ch. 3), is followed by an overview of the physics challenges and performance trade-offs of competing passive HF and UHF RFID systems by Marlin Mickle, Director, and colleagues Peter J. Hawrylak and Leonid Mats from the RFID Center of Excellence at the University of Pittsburgh (Ch. 4). Specifications for active RFID sensors and a proposal to standardize interfaces to active RFID sensors, building on the EPCglobal RFID and IEEE1451 sensor interface specifications, are introduced by Kang Lee of NIST and Tom Cain, Ph.D., University of Pittsburgh (Ch. 5). A test methodology for evaluating real-time location systems with RFID systems, starting with IEEE 802.11g and ISO 24730 Part 1 Real Time Locating Systems (RTLS), is introduced by Mohammad Heidari and Kaveh Pahlavan, Director of the Center for Wireless Information Network Studies at Worcester Polytechnic Institute (Ch. 6). A simulation methodology for modeling the EPC network is presented by John Williams, Director, and Abel Sanchez, Ph.D., of the MIT Auto-ID Labs and colleagues from SAP Research (Ch. 7). In the conclusion we will revisit the question of how passive RFID technology for the supply chain integrates with sensor networking and location tracking, and how these applications complement and/or conflict with current RF infrastructure and applications from aerospace to medical and retail facilities.

In the RFID applications section of this book (Chs. 8–14) Giselle Bennett, Director, Logistics and Maintenance Applied Research Center, and Ralph Herkert of the Georgia Tech Research Institute at Georgia Institute of Technology expose the challenges of deploying active RFID systems (Ch. 8) from their experience managing projects for the US Navy. Bill Hardgrave, Director of RFID Research Center at the University of Arkansas and Robert Miller, Ph.D., of the Dauch College of Business at Ashland University in Ohio, follow with their assessment of challenges and opportunities for achieving visibility in cross-border international supply chains (Ch. 9). Duncan McFarlane, Director of the Cambridge University Auto-ID Labs and colleagues Alan Thorne, Mark Harrison, Ph.D., and Victor Prodonoff Jr. describe creating an Aero-ID Programme research consortium with the largest US and European exporters who are using RFID identification and tracking technology in restructuring the aerospace industry supply base (Ch. 10). J. P. Emond, Co-Director of the Center for Food Distribution and Retailing, shares the challenges of using temperature sensors with RFID tags in "cold chain" applications for fresh produce and pharmaceuticals (Ch. 11). Thorsten Staake from the Auto-ID Labs at MIT with colleague Florian Michahelles, and Elgar Fleish, Director of the Auto-ID Labs at St. Gallen, address RFID technology and application problems in anti-counterfeiting (Ch. 12).

When RFID data is shared in the context of new business processes that extend beyond the four walls of the enterprise, entirely new possibilities for visibility emerge. Dimitris Kiritsis describes (Ch. 13) how RFID systems are being designed at Fiat to track products across their lifecycles for managing vehicle disposal, which today accounts for 10% of hazardous waste in landfills in Europe, and how W3C Semantic Web technology can be used to link uniquely identified objects with conceptual objects such as processes, agents, and time stamps using the RDF representation scheme. The RFID applications chapters close with an exploration of autonomic logistics by Bernd Scholz-Reiter and Dieter Uckelmann with researchers Christian Gorldt, Uwe Hinrichs, and Jan Topi Tervo of the Division of Intelligent Production and Logistics Systems, University of Bremen (Ch. 15).

Throughout each chapter we explore how RFID may be used to unlock information from manual entry or "line-of-sight" barcode data acquisition scanning processes, as well as from proprietary enterprise data models, to enable cross-company, cross-industry, and cross-country information services about products, their condition, and where they are. "How does the world change," observes Hao Min, Director of the Auto-ID Lab at Fudan University in Shanghai while working on his contribution to this book, "when the 'Internet of things' contains a profile for every object. If we contrast this to the internet today, information about people and events are recorded and Google is used to search information about people and events. If the information (profile) of every object (include people) is recorded, what will the internet be like?"

The market for RFID technology is growing rapidly, with significant opportunities to add value, but also, because of the challenging engineering issues that are identified in this book, many opportunities for failure. At the 2007 Smart Labels Conference here in Boston, Raghu Das, CEO of IDTechEx, estimated that almost half as many tags will be sold this year as the total cumulative sales of RFID tags for the prior sixty years of 2.4 billion. While approximately 600 million tags were sold in 2005, expectations for 2006 are for sales of 1.3 billion tags in a $2.71 billion market. Of that amount "about 500 million RFID smart labels will be used for pallet and case level tagging but the majority will be used for a range of diverse markets from baggage and passports to contactless payment cards and drugs."[2] The total market for passive RFID tags conforming to international

---

[2] RFID Smart Labels 2007 – IDTechEx, February 20–23, 2007, Boston Marriott, Boston, MA (http://rfid.idtechex.com/rfidusa07/en/RFIDspeakers.asp).

interoperability standards for supply chain applications has not yet been growing as quickly as anticipated. The challenge of gaining market share for any disruptive technology in an established market such as RFID requires selling much higher volumes of low-cost items to impact industry sales, as is the case for passive UHF RFID tags that are priced at under 15 US cents.

In fact, not only is the overall number of RFID tags being sold doubling, but also the numbers of technology choices are expanding rapidly. RFID transponders (receiver–transmitter "tags") as part of a class of low-cost sensors are evolving to include more or less intelligence (processors, memory, embedded sensors) on a variety of platforms (from semiconductor inlays or MEMs to inorganic and organic materials that form thin film transistor circuits – TFTCs) across a variety of frequencies (UHF, HF, LF) and protocols (802.11, Bluetooth, Zigbee, EPC GenII/ISO 18000-7). One of the resulting challenges for planning RFID systems is the necessity to keep track of the evolving technology, from semiconductor inlays and printed antenna designs for RFID tags, both passive and active, via high-speed applicators and reader engineering, to sensor networking and the definition of new shared business processes.

Somewhere between an overall RFID market that promises to deliver more tags in the next year than in the prior sixty years since RFID was invented and specific industry sectors where penetration of RFID and, in some cases, even barcode usage is low, there are significant opportunities to use RFID for improving efficiency and visibility. The breakthrough in low-cost RFID tags for everyday products has occurred as a result of the adoption of specifications for interoperable UHF RFID tags that were developed by and licensed from the MIT Auto-ID Center by the barcode associations, now known as GS-1, and the nonprofit EPCglobal industry membership consortium that was formed to promote the use of RFID in today's fast-moving, information-rich, trading networks. The longer read range (several feet at a minimum) requirements for loading dock and warehousing applications, as well as recent UHF near-field research for closer-range applications as presented in this book, make the EPC GenII/ISO 18000-6C specifications a leading contender for passive RFID systems where global interoperability is required. This is clearly the case for supply chains where tags on products manufactured in one country must be read by RFID interrogators (transmitter/receiver "readers") halfway around the world.

In addition to low-cost passive RFID technology, the authors explore active RFID technology for adding telemetry and real-time location system (RTLS) data. During the initial proposal review process with Cambridge University Press for this book, reviewers questioned the usefulness of adding sensor and RTLS technology to a field that, as Dr. Julie Lancashire, Engineering Publisher, describes, is so "massively multidisciplinary." Subsequently the market validated the importance of incorporating active RFID technology for asset tracking and condition-based monitoring applications with the recent acquisitions of Savi Networks by Lockheed Martin and of Wherenet by Zebra Technologies, whereby

RTLS systems now account for up to 30% of RFID systems sold, according to the IDTechEx market study cited above. As a matter of scope, this initial book does not explore RFID systems that work at close (several inches) proximity that are being deployed for access control, personal and animal identification, and payment processing systems based on emerging standards such as IEEE 802.15.4 WPAN and NFC

The chapters that follow address these opportunities from the perspective of principal researchers who have been engaged in the RFID Academic Convocations with senior executives from "first mover" market-leading companies in aerospace and healthcare life sciences, as well as from retail "cold chain" and fast-moving consumer goods supply chains. As Gerd Wolfram, Managing Director of Advanced Technologies at Metro Group Information Technology, said in his address to the EU RFID Forum 2007/4th RFID Academic Convocation,[3] the development of interoperability standards has truly been a community effort with input from academics, industry users, and service providers, as well as from non-governmental and government agencies around the world.

One industry that is establishing benchmarks for how RFID can be used for securing the supply chain and is working to harmonize compliance reporting across jurisdictions is healthcare. At the 5th RFID Academic Convocation pre-conference co-hosted by the Auto-ID Labs and RFID Live 2007 in Orlando,[4] Ron Bone, Senior Vice President of Distribution Planning for McKesson Corporation, and Mike Rose, Vice President RFID/EPCglobal Value Chain for Johnson and Johnson, who serve as EPCglobal Healthcare Life Sciences (HLS) Business Action Group Co-Chairs, spoke about the industry's progress in working proactively with government agencies for a safer and more secure pharmaceutical supply chain. At this gathering the Office of Science and Engineering Labs Center for Devices of Radiological Health at the US Food and Drug Administration (FDA) also presented findings and discussed test methodologies for evaluating the impact of RFID on medical devices. In evaluating EPC network components, from RFID tags to network registries, a common theme emerged from the academic papers that were presented, of an ongoing requirement for fact-based simulation and test methodologies to evaluate various RFID scenarios under consideration, an approach that is pursued in the chapters that follow.

Carolyn Walton, Vice President of Information Systems for Wal-Mart, stated in her address at the 5th RFID Convocation that healthcare costs are growing more quickly than company profits and threaten to overcome national healthcare programs such as Medicare. Citing the $11 billion in excess costs identified by the Healthcare Information and Management Systems Society (HIMMS) study of

---

[3] EUROPE RFID Forum 2007, organized in conjunction with the 4th RFID Academic Convocation in Brussels (http://ec.europa.eu/information_society/policy/rfid/conference2007_reg/index_en.htm).

[4] 5th RFID Academic Convocation – Pre-conference to RFID Live! April 30, 2007, Disney Coronado Springs Resort, Orlando, FL (http://www.rfidjournalevents.com/live/preconfer_academic_convocation.php).

hospitals,[5] where supplies consume 40% of operating costs and administration exceeds 25% of supply chain costs, and especially where barcode technology has not been implemented, Carolyn said that there is an opportunity for supply chain management best practices including RFID.

RFID together with wireless sensors and actuators are extending the reach of the internet in ways that promise to transform our ability to communicate about, and interact with, things in the physical world. These chapters are written from the very different perspectives of principal investigators from their diverse areas of research interest. Nonetheless, the authors share an interest in and vision of RFID technology that facilitates communication and enables better visibility and management decisions. The challenge that we face – and would like to invite readers of this book to explore – is one of finding out how we can combine data related to unique IDs to create applications that add value to our communities and to commerce. An acknowledgment of individuals who have supported this collaboration follows this preface.

Stephen Miles, Sanjay Sarma, and John R. Williams
Massachusetts Institute of Technology, Auto-ID Labs,
Cambridge, MA

---

[5] The Healthcare Information and Management Systems Society (HIMSS) breaks down savings into four categories: $2.3 billion inventory management, $5.8 billion order management, $1.8 billion transportation, and $1.1 billion physical distribution; 14th Annual HIMSS Leadership Survey.

Acknowledgments

This book is the direct result of the research collaboration initiated at the RFID Academic Convocation in January of 2006 hosted by the Auto-ID Labs at MIT that was organized with the support of co-editors Sanjay Sarma, co-founder of the Auto-ID Center, and John Williams, Director of the Auto-ID Labs at MIT, and that subsequently evolved to include events in Shanghai and Brussels with co-sponsorship from the Chinese Academy of Sciences and the Ministry of Science and Technology (MOST) and the European Commission Directorate General for Informatics (DGIT). I would like to thank the authors included in this book who served as Conference Committee members during this process for their input and support.

On behalf of the Conference Committee, special thanks go to industry leaders Simon Langford, Director at Wal-Mart, and Mike Rose, Vice President RFID/EPC Global Value Chain at Johnson and Johnson, with Ron Bone, Senior Vice President Distribution Planning at McKesson, for taking the time from their busy schedules and leadership roles within the EPCglobal Healthcare Life Sciences community to investigate issues requiring broader research collaboration. I would like to personally acknowledge Convocation Co-Chairs Bill Hardgrave, Director of the RFID Lab at the University of Arkansas, and John Williams, Director of the Auto-ID Labs at MIT, as well as Co-Sponsors Yu Liu, Deputy Director of the RFID Laboratory of the Institute for Automation at the Chinese Academy of Sciences, Dr. Zhang Zhiweng, Department of High-Technology Development & Industrialization of the Chinese Ministry of Science and Technology, and Hao Min, Director of the Auto-ID Lab at Fudan University, for their support.

From the EU RFID Forum/4th RFID Academic Convocation I would like to acknowledge Organizing Committee members Henri Barthel, Director of the European Bridge Project and Technical Director of EPCglobal Europe, Duncan McFarlane, Program Committee Chair and Director of the Auto-ID Lab at Cambridge University, Dimitris Kiritsis of the EPFL, and Co-Sponsors Peter Friess, Florent Frederix, and Gerard Santucci, Head of the European Commission Directorate General for Informatics (DGIT), Networked Enterprise & Radio Frequency Identification (RFID), for their leadership. Finally the Auto-ID Labs have benefited from the continued engagement of the original Board of Overseers of the MIT Auto-ID Center and current EPCglobal Board of Governors members who are engaged in Asia and Europe as well as the

Americas. Several of these executives, including Dick Cantwell, Vice President of Procter and Gamble, who serves as Chairman, and Sanjay Sarma, co-founder of the MIT Auto-ID Center and co-editor of this book, continue to serve on the EPCglobal Board of Governors and to provide examples of "RFID advantaged" applications.

It has already been three years since the Auto-ID Labs were formed, together with the creation of EPCglobal and their commitment to fund academic research to support the RFID technology licensed from the MIT Auto-ID Center. At the MIT Auto-ID Labs inception, I organized the Auto-ID Network Research Special Interest Group (SIG) to investigate requirements for Electronic Product Code (EPC) data exchange, with principal investigators who served in succession, co-founders of the Auto-ID Center Sanjay Sarma, David Brock, Ph.D., and current MIT Auto-ID Labs Director John Williams. We would like to acknowledge sponsors Jim Nobel, CIO of the Altria Group, and his Global Information Services team leaders Stephen Davy and Brian Schulte, Tom Gibbs, Director of Global Solutions at Intel, Ajay Ramachandran, CTO of Raining Data, and David Husak, CTO of Reva Systems, for their support of a sponsored research initiative to use web protocols for communicating about things. CIO Ramji Al Noor and Steve Stokols in their roles at Quest (prior to transitioning to British Telecom), Matt Bross, CTO of British Telecom, with Peter Eisenegger, Steve Corley, and Trevor Burbridge of the R&D group in Martlesham Health, and Dale Moberg, Chief Architect of Cyclone Commerce (now Axway), were instrumental in validating the opportunity for creating Auto-ID/RFID services, as was Alan Haberman, a father of the barcode movement and an early instigator of this research initiative. Special thanks are due to Tim Berners-Lee and to Steve Bratt from W3C for their continued support and vision of a world where information can be retrieved and re-used in ways that had not been envisioned when it was created.

From the Management of Technology (MoT) Program at the MIT Sloan School I would like to thank MoT Program Director Jim Utterback for reviewing an early version of my chapter and for his teachings that bring an historical perspective to disruptive technologies – including the rediscovery of a nineteenth-century export trade in ice blocks based in Boston – and Tom Eisenmann of the Harvard Business School for his best practices case studies in "Riding the Internet Fast Track," whose company founders he brought into the lecture hall to present their case studies to classmates who had survived the dot.com "bubble."

I would also like to express my gratitude to colleagues who worked with me in designing internet solutions for data communications including VoIP, MPLS core routers, and mobile IP telephony at Officenet, NMS Communications, Ironbridge Networks, and Wireless IP Networks, respectively, as we worked the world over to deploy infrastructure for adding value through IP networks. Special thanks go to my family Ingrid, Garth, and Stephen, who grew up in the turbulent world of high-technology businesses. It is my hope that this book will pave the way for people to use Auto-ID technology to apprehend and actuate better decisions for our work, our entertainment, and the world environment in which each one of us plays a unique role.

Stephen Miles
Auto-ID Labs, Massachusetts Institute of Technology,
Cambridge, MA

1 Introduction to RFID history and markets

Stephen Miles

The market for radio frequency identification (RFID) technology is growing rapidly, with significant opportunities to add value, but also, because of the challenging issues that are identified in the chapters that follow, many opportunities for failure. This book brings together pioneering RFID academic research principals to analyze engineering issues that have hampered the deployment of RFID and to share "best practices" learnings from their work, building on the tradition of the Auto-ID Labs. The Auto-ID Labs consortium of leading universities around the world includes Auto-ID Labs at Cambridge University, Fudan University, Keio University, the University at St. Gallen and the ETH Zürich, the University at Adelaide and, most recently, the ICU, South Korea.[1] The principal investigators represented here have conceived, obtained funding for, and executed research projects using RFID technology. The authors share their experience in the design, test, prototyping, and piloting of RFID systems, both to help others avoid "reinventing the wheel" and to set the stage for what is next in RFID.

Because RFID technology has evolved from proprietary systems operating at different frequencies in jurisdictions with different RF regulatory restrictions, most RFID work has been divided into communities operating at one frequency or another. In *RFID Technology and Applications* we bring together principal investigators with experience in passive RFID systems across a range of frequencies including UHF 860–960 MHz (EPC GenII/ISO 18000–6c) and HF 13.56 MHz (ISO 18000-3),[2] but also, breaking with precedent, we include experts

---

[1] The Auto-ID Labs are the leading global network of academic research laboratories in the field of networked RFID. The labs comprise seven of the world's most renowned research universities located on four different continents (www.autoidlabs.org).

[2] ISO specification for RFID under the standard 18000-1 Part 1 – Generic Parameters for the Air Interface for Globally Accepted Frequencies at frequencies per below can be obtained from http://www.iso.org/iso/en/CombinedQueryResult.

18000-2 Part 2 – Parameters for Air Interface Communications below 135 kHz
18000-3 Part 3 – Parameters for Air Interface Communications at 13.56 MHz
18000-4 Part 4 – Parameters for Air Interface Communications at 2.45 GHz
18000-5 Part 5 – Parameters for Air Interface Communications at 5.8 GHz (withdrawn)

RFID Technology and Applications, eds. Stephen B. Miles, Sanjay E. Sharma, and John R. Williams. Published by Cambridge University Press. © Cambridge University Press 2008.

in active (with power) RFID systems. The inclusion of active RFID with passive systems allows us to explore a wider range of technologies for how one might best add "real-world awareness," such as location and sensor data, to the information about identified objects – both of which are high-growth markets, as cited in the Preface. Researchers gathered in this collection of essays have been selected for their experience as principal investigators and RFID lab directors and from their participation in the RFID Academic Convocations that are being held around the world with industry and government leaders to explore issues requiring greater research collaboration.[3]

This chapter provides an historical introduction to RFID together with an overview of the standards and regulatory frameworks that cross frequencies, protocols, and processes to govern how we engineer RFID systems to operate in different jurisdictions. Recent breakthroughs in global standards and regulatory initiatives in European and Asian countries have freed unlicensed UHF radio spectrum for use by RFID systems, making this a seminal moment to examine new system design possibilities. During the spring of 2007 the conditional approval of UHF as well as HF frequencies for RFID applications in China and Europe, the adoption of a variety of technical standards for passive and active RFID systems into the International Standards Organization (ISO) process, the availability of much of this technology under Reasonable and Non Discriminatory Licensing (RAND – see Section 15.7) terms, and the release of the Electronic Product Code Information Services (EPCIS) software specifications for exchanging data about products from EPCglobal all promise to make it possible to communicate more effectively about the condition and location of products. The chapters that follow explore the underlying technology and growing markets for asset-tracking and cold-chain and condition-based monitoring across entire supply chains and product lifecycles.

The technology chapters begin with a deep dive into the design of low-power passive and active RFID transponders (tags) and RF performance in near-field and far-field modes over HF and UHF frequencies. The "Swiss cheese effect" of RF "null" zones caused by multipath effects, as well as "ghost tags," the bane of indoor RF systems, are introduced by Hao Min, Director of the Auto-ID Lab at Fudan University in Shanghai (Ch. 2), together with recommendations for addressing these issues at a tag and, in subsequent chapters, at the system and supply chain network level. The RFID applications chapters of this book present hands-on research experience of principal investigators in specific markets and illustrate the promise they see for changing the way businesses is done. "How does the world change," observes Hao Min while working on his contribution to this

---

18000-6 Part 6 – Parameters for Air Interface Communications at 860 to 960 MHz
18000-7 Part 7 – Parameters for Air Interface Communications at 433 MHz
[3] The RFID Academic Convocation co-hosted by the Auto-ID Lab at MIT brings together RFID research principals, leaders from industry and government, and technology providers to address research issues surrounding the implementation of RFID (http://autoid.mit.edu/CS/blogs/convocations/default.aspx).

book, "when the 'Internet of things' contains a profile for every object. If we contrast this to the internet today, information about people and events are recorded and Google is used to search information about people and events. If the information (profile) of every object (include people) is recorded, what will the internet be like?"

One of the first issues RFID project managers face is the lack of diagnostic tools to characterize RF environments and RF tag performance on specific products, short of working within a fully instrumented anechoic chamber and using finite state analysis to simulate indoor RF propagation fields. As Larry Bodony, one of the members of the MIT Enterprise Forum RFID Special Interest Group commented recently on his experience implementing RFID container tracking systems with Lockheed Martin/Savi Networks, "RFID is like an Ouija Board, where you address one RF problem and another issue pops up somewhere else" [1].

Like the Ouija Board, RFID's roots can be traced to the period of spiritualist practices in the mid nineteenth century and to Maxwell, who first predicted the existence of electromagnetic waves. In the chapters that follow we ask the following question: "what are the RFID engineering issues that need to be addressed to fulfill the promise of increased visibility and collaboration?" As an organizing principle for the chapters that follow, control systems methodology is introduced as an approach to addressing RFID engineering issues. The theme of control systems methodology for systems design spans recent work by co-editors Sanjay Sarma, co-founder of the Auto-ID Center, on "Six Sigma Supply Chains" [2], as well as by John Williams, Director of the Auto-ID Labs at MIT, on "Modeling Supply Chain Network Traffic" (Ch. 7).

1.1   Market assessment

The sales of passive RFID UHF tags conforming to international interoperability standards for supply chain applications are not yet growing as quickly as anticipated, despite, or perhaps because of, their low price – under 15¢ for EPC GenII/ ISO 18000-6c transponders (tags). At the 2007 Smart Labels Conference in Boston, Raghu Das, CEO of IDTechEx, estimated that sales of RFID tags for 2006 would total 1.3 billion tags. As cited in the preface "about 500 million RFID smart labels will be used for pallet and case level tagging but the majority will be used for a range of diverse markets from baggage and passports to contactless payment cards and drugs."[4]

The overall number of RFID tag unit sales is estimated to grow by 100% this year, in conjunction with costs that are dropping by 20% or more per year. The range of technologies is also growing at a rapid pace, both for low-cost and for

---

[4] RFID Smart Labels 2007 – IDTechEx, February 20–23, 2007, Boston Marriott, Boston, MA (http:// rfid.idtechex.com/rfidusa07/en/RFIDspeakers.asp).

high-end tag functionality. RFID tags as a class of low-cost sensors are evolving to include more or less additional intelligence (processors, memory, embedded sensors) on a variety of platforms (from semiconductor inlays and MEMs to inorganic and organic materials that form thin film transistor circuits – TFTCs) across a variety of frequencies (UHF, HF, LF) and protocols (802.11, Bluetooth, Zigbee, EPC GenII).

In meetings with industry stakeholders, government representatives, and academic researchers in search of better ways to communicate about products at the RFID Academic Convocations, one is struck by the diversity of applications for RFID. Researchers have discussed applications from tracking the condition of containers across oceans to production control in semiconductor "lights out" factories, from keeping tabs on draft beer kegs for pubs in Britain to monitoring sales promotions compliance across retail distribution networks.

In spite of the overall market growth for tagging products, it is still difficult to communicate outside "the four walls" of an enterprise today. Most RFID systems have been "closed loop" proprietary technologies, while the promise of enhanced supply chain visibility and eBusiness collaboration using low-cost "open loop" interoperable passive UHF RFID tags is still hampered by technical issues as are explored in depth herewith. While the passive UHF RFID tags envisioned at MIT were restricted by design to minimize cost and therefore contained just enough memory to store a unique identifier the length of an EPC code, RFID tag manufacturers continue to develop a wide range of systems operating at various frequencies and combinations thereof. Suppliers are also adding memory and processing power for storing and processing additional information about the condition, maintenance history, and/or location of individually tagged assets. We will learn from the authors how far RFID technology has come in these applications, in an overall market that promises to deliver more tags in the next year than in the prior sixty years since RFID was invented, and what new applications they are exploring that may change the world.

1.2   Historical background

The history of radio frequency engineering can be traced to 1864 when James Clerk Maxwell predicted the existence of electromagnetic waves, of which microwaves are a part, through Maxwell's equations. By 1888, Heinrich Hertz had demonstrated the existence of electromagnetic waves by building an apparatus that produced and detected microwaves in the UHF region, the radio frequency selected by the Auto-ID Center at MIT for its passive RFID initiative a century and a half later. From Dr. Jeremy Landt's *Shrouds of Time; the History of RFID*, we learn that "[Maxwell's] design used horse-and-buggy materials, including a horse trough, a wrought iron point spark, Leyden jars, and a length of zinc gutter whose parabolic cross-section worked as a reflection antenna." [3].

Radio frequencies, like other physical signals in nature, are analog, as is also the case for voltage, current, pressure, temperature, and velocity. Radio frequency waves and radar radiation connect interrogators and tags via "inductive coupling" or "backscatter coupling" as will be analyzed in depth by Marlin Mickle in "Resolution and Integration of HF & UHF" (Ch. 4). The first RFID applications were developed in conjunction with radar technology at the height of the Second World War, for Identification Friend or Foe (IFF) systems, where the RF transponder (tag) and interrogator (reader) were designed to detect friendly airplanes. [4] A precursor to passive RFID was the electronic article surveillance (EAS) systems deployed in retail stores in the 1970s that used dedicated short-range communication (DSRC) RF technology for anti-theft detection.

Auto-ID RFID technology builds on automated data capture (AIDC) barcode standards for identifying products that, together with the standardization of shipping container dimensions, have so dramatically lowered the cost of transportation in recent decades [5]. Companies that seized the opportunity to optimize their supply chains with this technology have become some of the largest companies in the world, including such retailers as Wal-Mart, Metro, Target, and Carrefour. The best-known and most widespread use of AIDC barcode technology has been in consumer products, where the Universal Product Code (UPC) was developed in response to grocery industry requirements in the mid 1970s [6] [7]. Where barcodes are widely used in these networks today, RFID systems are now being installed to expedite non-line-of-sight data capture using RF to read the electronic product code (EPC) on RFID tags.

One area where history can help us to avoid "reinventing the wheel" is in human resources planning for what RF background is useful for implementing low-power RFID systems today. In my search for domain expertise in this area I was surprised to learn, upon meeting two leading RF experts, Peter Cole, Ph.D., the Director of the Auto-ID Lab at Adelaide University, whose work was instrumental in the initial MIT Auto-ID Center UHF specifications, and Marlin Mickle, Ph.D., the Nickolas A. DeCecco Professor and Director of the RFID Center of Excellence at the University of Pittsburgh, that they are both septuagenarians. They were students during the Second World War when radar technology was saving lives during the Blitz bombardments against London. Radar engineers are familiar with the multipath effects that cause ghost targets to appear, phenomena that haunt RFID data acquisition to this day. By contrast, the RF engineers who have worked in more recent RF domains such as cellular telephony and wireless LANs are accustomed to working at much higher power levels and with more host processing capabilities in cell phones than are present in tiny RFID tags. This insight was confirmed by SAAB executives[5] in a meeting at the MIT Auto-ID Labs (April 10, 2007), who related their success in bringing radar technicians from the airplane manufacturing side of the business to design their first EPC GenII/ISO 18000-6c RFID infrastructure.

---

[5] Goran Carlqvist, Lars Bengtsson, and Mats Junvikat.

1.3   Adoption of the Auto-ID system for the Electronic Product Code (EPC)

I have the good fortune of writing this introduction from the perspective of the MIT Auto-ID Labs, successor to the Auto-ID Center, where, with input from sponsors Gillette, Proctor and Gamble, and the Uniform Code Council (now GS-1), the specifications for a passive UHF RFID Electronic Product Code system were conceived and where current research is expanding the boundaries of the "internet of things." In the 2001 white paper *The Networked Physical World, Proposal for Engineering the Next Generation of Computing, Commerce and Automatic Identification*, co-authors Sanjay Sarma, Ph.D., David Brock, Ph.D., and Kevin Ashton [8] proposed a system for the Electronic Product Code. The name for the language to communicate the whereabouts of an object in time and place was the Physical Markup Language (PML) [9]. The specifications for UHF passive tags and RFID interrogators developed at the Auto-ID Center were subsequently licensed by EPCglobal, a standards body that was formed from the article-numbering barcode associations around the world, to promote the use of RFID in commerce.

As Alan Haberman, an early proponent of barcodes who served on the Board of Governors for the new EPCglobal association, reminded us in his review of this introductory chapter, the nature of the technology transfer process for the Auto-ID system from academia to industry differed from other technology licensing agreements. In this case, the licensor was GS-1, an organization that had grown from the adoption by the grocery industry of a standard for barcodes in 1973 – the Universal Product Code (UPC). Today GS-1 is a not-for-profit organization representing 1.2 million company users in 130 nations that develops, promotes, and governs, through the participation of its members, standards for automatic identification of product, location, and process worldwide. GS-1's investment in RFID reflects their view of the sea change in technology for identification that RFID represents and their commitment to supporting the maturation and technology transfer to industrial and commercial users around the globe. GS-1's commitment to financing the development of a new organization, EPCglobal, its investment in spreading the word and building that organization, and the initial five-year commitment of $2,000,000 per annum to the Auto-ID Labs consortium led by MIT for ongoing research are some of the elements that have positioned RFID technology for worldwide adoption.

In addition to the challenging physics of generating low-power electromagnetic UHF signals to wake up passive RFID tags to transmit their ID in the original MIT research, a major hurdle in establishing the Auto-ID system, whereby RFID readers in one country could read RFID tags from another, began with ensuring the availability of common unlicensed radio frequencies across national jurisdictions. Starting with frequency regulations, freeing up the 860–960 MHz UHF spectrum for EPC GenII/ISO-18000-6c RFID transmissions has presented a challenge, both in Europe, where regulatory limits to transmission power and "listen before talk" restrictions hampered UHF RFID performance, and in Asian countries, where these frequencies were licensed for other uses. A great number of individuals from industry and governmental organizations around the world have contributed to opening this unlicensed spectrum for the use of RFID, including the Research Directors of the Auto-ID Labs, who have been instrumental in their respective countries in establishing a constructive dialogue with government and industry on interoperable standards for RFID.

In May of 2004, following the formation of the Auto-ID Labs, David Brock, Ph.D., a co-founder of the Auto-ID Center, and I were invited to participate in the Forum on eBusiness Interoperability and Standardization organized by Dr. David Cheung, Director of the University of Hong Kong's Centre for E-Commerce Infrastructure Development (CECID).[6] One outcome was MIT Auto-ID Labs' support for the successful application by Ms. Anna Lin, President of the Hong Kong Article Numbering Association, for a \$1.8M RFID pilot grant to track goods manufactured in southern China's Pearl River Delta funded by the Hong Kong Innovation & Technology Commission [10]. Building on the success of this project, the Japanese Ministry of Economics, Trade, and Industry (METI) is sponsoring cross-border RFID pilots between China and Japan. One of the early research participants, Dr. S. K. Kwok, Project Fellow of the Hong Kong Polytechnic University, presented his research findings in "RFID for Enhancing Shipment Consolidation Processes" at the RFID Academic Convocation in Shanghai [11]. The Auto-ID Labs at MIT support international standards through hosting groups, such as a recent visit by the Coordination Program of Science and Technology Project of the Japan Science and Technology Agency, as well as leading academic researchers in healthcare and logistics and advisors to government agencies on technology strategy. One recurrent theme that we hear from different national technology planners is how much better off the world is as a result of achieving consensus around IEEE 802.11 WiFi protocols for unlicensed RF data communications, and how the EPCglobal GenII/ISO-18000-6c specifications for passive UHF RFID systems have similar promise.

Through hosting the RFID Academic Convocations noteworthy exchanges with the Auto-ID Labs have included Dr. Zhang Zhiwen of the Ministry of Science and Technology (MOST) and the RFID Lab at the Chinese Academy of Sciences (CASIA) in Beijing,[7] whose initial visit to MIT in October of 2005 led to his presentation of the Chinese Public Service Infrastructure project at the first RFID Academic Convocation (Cambridge, MA, January 23–4, 2006) [12]. Subsequent research collaboration sponsored by SAP Research using ISO 18000-6c systems in a secure supply chain is being explored both at the MIT Auto-ID Labs (Ch. 7) and directly with the RFID Laboratory at CASIA and with Haier Corporation, the world's biggest volume producer of white goods, in Qingdao,

---

[6] Forum on e-Business Interoperability and Standardization, organized with Jon Bosak, chair of the OASIS Universal Business Language Technical Committee (May 14, 2004, Hong Kong University, Hong Kong) (http://www.cecid.hku.hk/forum/file/ebusiness-eng.doc).
[7] http://www.rfidinfo.com.cn.

Shandong Province, People's Republic of China. The China Ministry of Science and Technology and Chinese Academy of Sciences (CASIA), together with Hao Min, Director of the Auto-ID Labs at Fudan University, co-sponsored the third RFID Academic Convocation (Shanghai, October 29–30, 2006) [13].

The participation by Dr. Peter Friess of the European Commission DGIT, with his contribution on "Networked Enterprise & Radio Frequency Identification (RFID)," in the first RFID Academic Convocation at MIT was equally noteworthy in that it led to the EU's participation in the Shanghai Convocation and subsequently to its hosting the RFID FORUM EUROPE 2007/RFID Academic Convocation (Brussels, March 13–14, 2007). Henri Barthel, Technical Director of GS-1 Europe and Coordinator of the EU-funded Bridge RFID Projects, led the Organizing Committee, together with Duncan McFarlane, Director of the Auto-ID Labs at Cambridge University, who also served as Programme Committee Chair with Dimitris Kiritsis of the EPFL.[8] At this event a representative of the European Telecommunications Standards Institute's Task Group 34 (ETSI TG34) presented the latest regulatory amendments to open additional channels for the use of ISO 18000-6c RFID devices in Europe. Robert Cresanti, Under Secretary of Commerce for Technology and Chief Privacy Officer, United States Department of Commerce, presented the US administration's support for industry-led interoperability initiatives subject to establishing security and privacy for consumers.

Following on the events described above and the collaboration of companies, industry groups, government agencies, and standards bodies, recent progress has occurred in defining specifications for RFID interoperability, most notably in the incorporation of EPCglobal GenII specifications for passive UHF RFID systems into ISO 18000-6c. China's State Radio Regulation Committee (SRRC) has conditionally approved the use of the 920.5–924.5 MHz and 840.5–844.5 MHz UHF spectrum for RFID systems [14]. Concurrently, at the US–European Commission Summit, President Bush and Chancellor Merkel announced an agreement to harmonize technology standards and best practices for RFID [15].

1.4   EPC information services

While RF frequencies and telemetry capabilities extend beyond the EPCglobal Electronic Product Code (EPC) system licensed from MIT, we make a point of considering, in the chapters authored by co-editors John Williams and Sanjay Sarma, the EPC Information Services (EPCIS) specification, released in April of 2007, for describing uniquely identified products in supply chain applications.

---

[8] The EU RFID Forum 2007 and the RFID Academic Convocation, co-hosted by the Auto-ID Labs at MIT, are being organized around the world to build collaboration across academic disciplines and institutional and geographic boundaries (Brussels, 13–14 March, 2007) (http://europa.eu.int/information_society/newsroom/cf/itemshortdetail.cfm?item_id=3132).

These specifications, developed in the EPCglobal Software Action Group and based on input from hundreds of EPCglobal member companies, represent an opportunity for computer systems to exchange machine-readable data about uniquely identified products, something that is not possible with today's web protocols. The EPCIS interface, for example, could be used for exchanging data independently of specific radio frequencies, transport protocols or even of RFID as a carrier (i.e. EPCIS could be used to exchange 2D barcode or active tag container tracking information). As EPCIS updates are released, the Auto-ID Laboratory at MIT is developing software documentation and data validation tools, together with an open source EPCIS code base, to enable rapid prototyping and testing of EPCIS data exchange over a variety of network bindings and programming paradigms.[9]

1.5   Methodology – closing the loop

Given the multidisciplinary nature of RFID applications and the complexity both of RF performance and of data exchange parameters for RFID-associated data, we propose the use of control theory as an experimental framework for RFID projects, whereby we define a use case, model the system, input data, and analyze the outcome to determine future inputs. This methodology can be used not only for quantifying tag RF power sensitivity (Ch. 3) or the accuracy of various location tracking algorithms (Ch. 5), but also for analyzing communications systems' performance and, at a business process level, measuring the inputs and outputs at various steps in a supply chain control loop. Control theory is at the heart of Six Sigma project management methodology as featured in Sanjay Sarma's work on "Six Sigma Supply Chains" (Ch. 2) that identifies the following processes: Define – Measure – Analyze – Improve – and Control. Control theory is also explored as a modeling framework for EPC network supply chain simulations by John Williams (Ch. 7).

An introduction to RF control systems modeling can be found in MIT Open-Courseware 6.661 taught by David H. Staelin, Ph.D., *Receivers, Antennas, and Signals* [16], that proposes a generic model for communications and sensing systems and also illustrates how passive RF sensing systems differ from active sensing and communications systems.

From Fig. 1.1 we can see how to (A) generate signals, which are processed (B), and then coupled to an electromagnetic environment (D) by a transducer or antenna (C). Another transducer (E) receives those electromagnetic signals and converts them into voltages and currents. In the case of passive RF systems, the transponder (tag) reflects information to the interrogator (reader). One challenge of passive RFID systems that is apparent from this diagram is that, on extending the range of the network to autonomous objects that are not connected to a

---

[9] The goal of EPCIS is to enable disparate applications to leverage electronic product code (EPC) data via EPC-related data sharing, both within and across enterprises; software tools are available at http://epcis.mit.edu/.

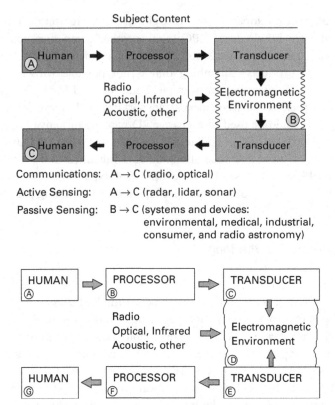

Fig. 1.1   Receivers, antennas, and signals (by David H. Staelin, from MIT OpenCourseware
6.661, 2007, with permission).

network, the information about those objects Ⓔ is not directly connected to an
originator "active" or "human" communicator Ⓐ, say, for example, like in the case
of a telephone conversation or a client/server or peer-to-peer data exchange. In
designing the architecture for an "internet of things" as is addressed in John Williams'
"EPC Network Simulation" (Ch. 7), reconstructing the business and security
context for "open loop" communications that register isolated EPC events and
create business events from them, using a specification such as EPCIS, remains a
challenging research area. The chapters that follow can be viewed as control loop
experiments for evaluating the use of different technologies, whereby a system is
modeled, and events are recorded and analyzed with respect to some metric for
measuring outcomes that in turn provides feedback for planning future projects.

1.6   RFID investing in a better future

Changes in regulatory and standards frameworks for RFID across many different
countries make this a seminal moment to evaluate RFID technology and

applications. In the fall 2006 Auto-ID Labs Research Meeting at the ICU in South Korea, Jun Murai, Director of the Auto-ID Labs at Keiyo, spoke in his keynote address about his interest in using RFID technology to address the environmental, educational, and welfare challenges of globalizing economies. In Asia, governments, most notably in China, Japan, and South Korea, have invested tens of millions of dollars in RFID research, to support national economic and public health priorities. South Korean government grants have been focused on changing demographics and how RFID systems can support residents staying in their homes as they age. At the EUROPE RFID Forum/RFID Academic Convocation in the spring of 2007 Gerard Santucci, head of the European Commission's DGIT Radio Frequency Identification (RFID) group, presented technology initiatives from Viviane Reding, Member of the European Commission, that included a report on the Bridge project ("Building Radio Frequency IDentification Solutions for the Global Environment") giving the results of a three-year 7.5 million Euro project, as part of the European Union's Sixth Framework Programme for Research and Technological Development.[10]

During the same period in the United States, government investments in academic research have dropped in many areas.[11] Nonetheless, it must be noted that United States government investment in RFID research contracts awarded during the current administration to consulting companies has skyrocketed. The largest of these projects, the US Department of Homeland Security (DHS) "Smart Border Alliance" consortium to design and implement the United States Visitor and Immigrant Status Indicator Technology (US-VISIT) program, has resulted in the RFID-based ePassport program. Accenture was awarded the contract "to help develop and implement a new automated entry/exit system to be deployed at the nation's more than 400 air, land and sea ports of entry" which, at $10 billion, makes it one of the biggest-ever US federal IT contracts.[12]

Recent outbreaks of contamination of meat with mad cow disease and contaminated pet foods remind us of the importance of mechanisms to track products back to their source across national borders. At the Third RFID Academic Convocation in Shanghai, the RFID Lab of the Institute for Automation at the Chinese Academy of Sciences reported on pilots for tracking blood supplies across multiple provinces in China and for monitoring fresh vegetables on their way to Japan, and on active tag RFID applications for container tracking,

---

[10] The Sixth Framework Programme (FP6), European Commission; The Sixth Framework Programme covers activities in the field of research, technological development, and demonstration (RTD) for the period 2002 to 2006 (http://ec.europa.eu/research/fp6/index_en.cfm?p=00).

[11] For example, DARPA funding for university researchers in computer science fell from $214 million to $123 million from 2001 to 2004. Vinton Cerf and Harris N. Miller, "America Gasps for Breath in the R&D Marathon," *Wall Street Journal*, New York, July 27, 2005 (for subscribers: http://online.wsj.com/article/SB112243461503197115-email.html).

[12] The program is currently facing strategic, operational, and technological challenges according to a report by the Congressional Budget Office, "BORDER SECURITY, US-VISIT Program Faces Strategic, Operational, and Technological Challenges at Land Ports of Entry," Congressional Budget Office report (http://www.gao.gov/new.items/d07248.pdf).

transportation, and military logistics. In the United States, the Prescription Drug Marketing Act of 1987 (PDMA) aims to increase safeguards in the drug distribution system to prevent the introduction and retail sale of substandard, ineffective, or counterfeit drugs. The State of California has mandated an electronic pedigree for prescription pharmaceutical sales that goes into effect 2009 [17]. Governments in Italy, Belgium, and Japan are following suit. In this book John Williams, Abel Sanchez, and SAP Research colleagues present a model to evaluate how ePedigree tracking can best be accomplished (Ch. 7).

1.7   New business processes

Returning to historical precedents that we can learn from in evaluating RFID information, the importance of establishing new processes to analyze new data sources can be seen in the history of the adoption of radar technology. At the beginning of the Second World War the British made adept use of this new source of information to defend against the Blitzkrieg. As stated in one account of this period, "early British experiments at Orfordness on the detection of aircraft by radar were turned with remarkable speed into a highly effective defense system. An essential factor in this achievement was the development of an extensive system of communications, special display equipment and, importantly, new military procedures. It is instructive to compare the operational results in the UK with the Japanese attack on the US naval base at Pearl Harbor, where radar systems had also been installed but where no effective plan for using this new source of information had been worked out" [4]. At the site of one of the most extensive deployments of radar technology at the time, Pearl Harbor, information analysts were unable to filter critical information from "noise" and could only sit and watch the incoming air raids as they occurred.

Like recognizing friend or foe, backscatter radiation from a passive RFID tag can make visible what is otherwise invisible. Early Auto-ID Labs research topics included the use of algorithmic data based on sensor inputs to generate real-time expiration dates or real-time demand-based pricing. What other processes might RFID technology track over the total product lifecycle according to recent EC directives for environmental waste disposal of electronic equipment?[13] During his recent visit to MIT, Jeff Bezos, CEO of Amazon, suggested that it might one day be important to measure the carbon-denominated impact of a supply chain transaction [18]. The lean manufacturing principles that govern today's most advanced manufacturing organizations have been applied to supply chain planning in David Simchi Levy's work in *Designing and Managing the Supply Chain* [19] that supports companies seeking to respond to market "pull." As organizations search to make their supply chains more resilient, to use Yossi Sheffi's term

---

[13] http://ec.europa.eu/environment/waste/weee/index_en.htm.

from *The Resilient Enterprise: Overcoming Vulnerability for Competitive Advantage* [20], many extended supply chains of today may need to be restructured along lean manufacturing guidelines. How far the retail supply chain has come in its implementation of RFID is examined by Bill Hardgrave and Robert Miller (Ch. 9).

One of the key learnings to emerge from the application scenarios that follow is that new processes must often be created to harness RFID information, as proposed in early research at MIT in this area [21]. This theme emerged from the Technology Day@MIT that the Auto-ID Labs hosted for the EPCglobal Board of Governors,[14] where Daniel Roos, author of *The Machine that Changed the World* and founder of the Engineering Systems Design Laboratory at MIT, further described how companies were breaking apart the brittle centralized manufacturing models of the industrial revolution to design more flexible business processes and tighter supplier collaboration to meet local market conditions.

RFID research and project management requires a balance between physics and electrical engineering, between computer science and business insight, in answering the question as to how best RFID data can add value. A new type of sensor has little value in the absence of control algorithms that can extract and make use of the data it presents in a business process. As Jim Utterbach points out in *Mastering the Dynamics of Innovation*, new technologies often exist for decades before their true potential is discovered, as exemplified in the history of the Marconi Wireless Telegraph and Signal Company's point-to-point ship-to-shore communications precursor to broadcast radio [22]. Making the invisible visible[15] is an opportunity that RFID makes available. How this can best be accomplished, and in what application areas, is explored in the chapters that follow. The RFID Academic Convocations provide a forum in which your participation is invited: see http://autoid.mit.edu/CS/blogs/convocations/default.aspx.

1.8   References

[1] **Bodony, L., MIT Enterprise Forum RFID Special Interest Group**, "RFID in Healthcare Supply Chains," May 7, 2007 (http://www.mitforumcambridge.org/RFIDSIG.html).

[2] **Sarma, S.**, "Keynote Address: Operational Excellence and Optimizing Performance in the Supply Chain," April 3–4, 2007, 2007 DoD RFID Summit, Washington DC (http://www.dodrfidsummit.com/agenda.html).

[3] **Landt, J.**, *Shrouds of Time. The History of RFID* (2001) (http://www.aimglobal.org/technologies/RFID/resources/shrouds_of_time.pdf).

[4] **Brown, L.**, *A Radar History of World War II: Technical and Military Imperatives* (Institute of Physics Publishing, London, 2000).

---

[14] Technology Day@MIT – The Auto-ID Labs-hosted "teach in" for the EPCglobal Board of Governors Meeting (July 20, 2006).

[15] An example of an EPCIS compatible geo-location service can be found in the GeosEPC demonstration at the MIT Auto-ID Labs site (http://epcis.mit.edu/).

[5] **Levinson, M.**, *The Box: How the Shipping Container Made the World Smaller and the World Economy Bigger* (Princeton University Press, Princeton, NJ, 2006).

[6] **Haberman, A. L.**, *Twenty-Five Years Behind Bars: The Proceedings of the Twenty-fifth Anniversary of the U.P.C. at the Smithsonian Institution, September 30, 1999* (Harvard University Wertheim Committee, Cambridge, MA, 2001).

[7] **Brown, S. A.**, *Revolution at the Checkout Counter: The Explosion of the Bar Code* (Harvard University Wertheim Committee, Cambridge, MA, 1997).

[8] **Sarma, S.**, **Brock, D.**, and **Ashton, K.**, *The Networked Physical World, Proposal for Engineering the Next Generation of Computing, Commerce and Automatic Identification* (Auto-ID Center, Cambridge, MA, 2001) (autoid.mit.edu/whitepapers/MIT-AUTO-ID-WH-001.pdf).

[9] **Brock, D.**, *PML* (Auto-ID Center, Cambridge, MA, May 2001) (autoid.mit.edu/whitepapers/MIT-AUTOID-WH-004.pdf). As is evident from this paper and in subsequent interactions, David was interested in standardizing a far larger swath of data about the physical world than has been applied so far in the EPCglobal standardization process.

[10] **Violino, B.**, "Leveraging the Internet of Things, with Standards for Exchanging Information over the EPCglobal Network Being Finalized, the Vision of Using RFID to Track Goods in the Supply Chain Is About to Become a Reality. And It Will Change Business as We Know It," *RFID Journal*, April 2005 (http://www.rfidjournal.com/magazine/article/2000/3/221/).

[11] **Kwok, S. K.**, "RFID for Enhancing Shipment Consolidation Processes," *RFID Academic Convocation*, Shanghai, co-hosted by CASIA and Auto-ID Labs at Fudan and at MIT (http://autoid.mit.edu/ConvocationFiles/China%20Agenda[1].pdf).

[12] **Hiwen Zhang**, "The Future of EPC Networks and Chinese Public Infrastructure," *RFID Academic Convocation* (Cambridge, MA, January 23–24, 2006) (http://autoid.mit.edu/CS/convocation/presentation%20in%20MIT_Zhang.ppt).

[13] **Heng Qian**, "Chinese Food Traceability Standards and the Potential Benefits of RFID," *RFID Academic Convocation* (Shanghai, October 26–27, 2006) (http://autoid.mit.edu/convocation/2006_10_26_Shanghai/presentations/P1_5_Qian_Heng_chinese_traceability_standards.ppt) (http://autoid.mit.edu/CS/blogs/announcements/archive/2007/05/04/27853.aspx).

[14] **Swedberg, C.**, "China Approves Requirements for UHF Bandwidth," *RFID Journal*, May 16, 2007 (http://www.rfidjournal.com/article/articleview/3318/1/1/).

[15] **US–EU Summit**, "Framework for Advancing Transatlantic Economic Integration between the United States of America and the European Union," US Office of the Press Secretary (Washington, DC, April 30, 2007) (http://www.whitehouse.gov/news/releases/2007/04/20070430-4.html).

[16] **Staelin, D. H.**, *Receivers, Antennas, and Signals, an Electrical Engineering and Computer Science Subject* (MIT OpenCourseware 6.661, Cambridge, MA, Spring 2003) (http://ocw.mit.edu/OcwWeb/Electrical-Engineering-and-Computer-Science/6-661Spring2003/CourseHome/index.htm).

[17] "Pharmaceutical Wholesalers and Manufacturers: Licensing," Chapter 857, *State of California Legislative Digest SB 1307* (Sacramento, CA, 2004) (http://www.dca.ca.gov/legis/2004/pharmacy.htm).

[18] **Bezos, J.**, "Emerging Technologies from Around the World and How to Make Them Matter," *2006 Technology Review Conference* (MIT, September 25–27, 2007).

[19] **Simchi-Levi, D.**, **Kaminsky, P.**, and **Simchi-Levi, E.**, *Designing and Managing the Supply Chain* (McGraw-Hill, New York, 2003).

[20] **Sheffi, Y.**, *The Resilient Enterprise: Overcoming Vulnerability for Competitive Advantage* (MIT Press, Cambridge, MA, 2005).

[21] **Subirana, B.**, **Eckes, C.**, **Herman, G.**, **Sarma, S.**, and **Barrett, M.**, "Measuring the Impact of Information Technology on Value and Productivity using a Process-Based Approach: The Case of RFID Technologies," *Center for eBusiness Research Brief*, Vol. V, No. 3 (October 2004) (http://digital.mit.edu/research/Briefs/Subirana_RFID_ CaseStudy_FINAL.pdf).

[22] **Utterbach, J.**, *Mastering the Dynamics of Innovation* (Harvard Business School Press, Boston, MA, 1994).

2   RFID technology and its applications

Sanjay Sarma

While the origins of RFID lie more than 50 years in the past, passive RFID technology is actually only in its infancy. This might seem an odd statement given that other technologies which have had a comparable history – computers for example – are considered mature. What makes RFID different?

The reason is that RFID, perhaps more than other technologies, is a *systems technology* that transcends the reader and the tag. Readers and tags are rarely, if ever, used alone. They are components of much larger systems, some of which they augment, and many of which they fundamentally enable. There are many other components to the system in which RFID participates, and, for RFID to really blossom, every component of the system must blossom. So every new advance in, say, battery technology or networking will launch a new wave of creativity and invention in RFID. This will create new applications. These new applications will increase the demand for products, further subsidizing research, and thus laying the seeds for the next invention and the next wave. It is my firm belief that RFID is currently only in its first wave.

EPC technology, developed first by the Auto-ID Center and then by EPCglobal, probably represents the state-of-the-art of the first wave of RFID. Today, EPC tags are being deployed worldwide in thousands of sites and billions of EPC tags have been read. Passive EPC tags are being used for intra- and inter-company applications on a scale perhaps never seen before. I will use the EPC system to describe the first wave of RFID systems. I will then speculate on the innovative waves that might come in the future.

2.1   The first wave: the state of EPC technology

A history of the development of the EPC is available in [1]. A detailed description is available in [2], and it would be redundant to repeat detail here. To summarize, the EPC stack, as shown in Fig. 2.1, has several layers starting from the tag. Readers read the tags using an air-interface protocol like EPC GenII [3]. The host software then talks to the reader using a standard like LLRP [4] or ALE [5].

RFID Technology and Applications, eds. Stephen B. Miles, Sanjay E. Sharma, and John R. Williams. Published by Cambridge University Press. © Cambridge University Press 2008.

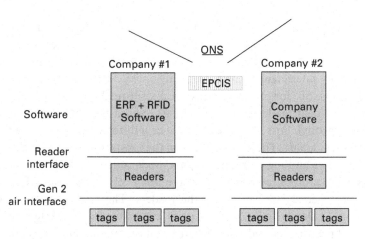

Fig. 2.1   The EPC stack.

Companies can then talk to each other using EPCIS [6]. Companies can "find" each other using ONS and extensions that are now being developed by EPCglobal.

RFID tags

A primary goal of the Auto-ID Center was to create an RFID technology of sufficient functionality that it is capable of supply chain applications, but yet inexpensive enough to be economically viable [7] [8]. Gen 1 EPC tags were the first step in that direction, but EPC GenII EPC tags seem to have achieved that goal. Published prices of commercially available inlays are already in the 5¢ range today [9].

The EPC GenII air interface

At the heart of the EPC suite of standards is the EPCglobal EPC GenII protocol for communication between readers and tags. The EPC GenII protocol is a very powerful one with a number of features almost unimaginable in a lower-cost tag even a few years ago. First, the EPC GenII tag permits fast read rates because of its innovative anti-collision algorithm and novel approach to putting tags to sleep. (Instead of putting tags to sleep once they have been counted, it puts tags from an A state into a B state and vice versa, which enables repeated reading without having to depend on a wake command.)

Second, the EPC GenII command set enables advanced sub-selection of tag populations. For example, it is possible for a reader to select only tags from Company A and tags from Company B for reading. This might seem esoteric but is actually quite useful in a number of situations – for example, reading P&G and Gillette tags at a P&G warehouse after P&G acquired the Gillette Company.

Third, EPC GenII gives a great deal of control over writing to tags, killing them, and locking them. It provides password control for each of these operations to be carried out with varying degrees of selectivity.

Finally, EPC GenII provides advanced features to address, and basically solve, a key problem in dense RFID: the problem of proximate readers interfering with each other's operation. EPC GenII tags have a number of safeguards and features that limit this interference. First, tags can operate with multiple readers in different sessions. So, for example, consider a shelf-reader that is reading its inventory. A hand-held reader can come along, request the shelf-reader to pause its operation (through host software), announce a different session number, read the tags, and then depart. The shelf-reader can then restart its reading where it left off. Second, readers and tags can operate in a mode called dense mode. In this mode, the tags, whose responses are, by definition, relatively weak, respond in bands that are not deafened by the signals of the readers. The reader–tag performance in dense mode is slower for individual pairs but better for the system. Finally, the EPC GenII protocol provides for different data rates and coding schemes – knobs which can be tweaked to adapt an RFID system to different circumstances.

Tag manufacture

The EPC GenII chip drives all EPC tags in use today. EPC GenII chips need to be attached to an antenna to work, and, though they are rather small, a number of techniques have been developed for attaching them to a substrate and to the antenna [10]. Antennas themselves are today manufactured largely by etching foils (mostly copper and aluminum), though techniques involving the deposition of copper, and deposition from conductive inks, are being studied in industry. A comprehensive description of the efficacies of these materials is available in [11]. A PET substrate with the chip and the antenna is called an *inlay*.

Unfortunately, currently additional steps are necessary to convert the tag from an inlay into a label because of business realities. Most EPC tags that are applied to a product today are applied to only a fraction of the total production. This is because only about 1,000 stores have been equipped with RFID, and there is no point wasting tags on products bound for the rest of the stores. As a result, tagging today is mostly done at DCs, and only for those products bound for the relevant stores. This *post facto* tagging strategy means that RFID tags need to be converted to adhesive labels, sometimes coated with anti-static materials, and then manually applied to the package. This is fairly expensive. Instead of the sub-10¢ prices of inlays, applied tag costs can be upwards of 20¢. This has skewed ROI computations and impeded adoption. The move from this approach to tag-at-source-type approaches, whereby the inlay is applied directly to packaging, will be swift and create a "virtuous cycle." Costs will drop, creating a better ROI. Adoption will increase, creating scale for manufacturers and offering the opportunity to offer lower prices.

Tag performance

There is a fundamental trade-off in the performance of RFID tags. In the low-frequency (LF) and high-frequency (HF) ranges, RFID readers tend to couple with RFID tags using the magnetic or electric fields. The magnetic field is more "penetrative" and more reliable in being able to power tags; however, the range of magnetic fields is relatively small – about the size of the antenna. In UHF, where the wavelength is lower, the electric and magnetic fields tend to couple by the time they reach the tag – in the form of electromagnetic waves. The range of UHF tags is much greater, sometimes on the order of several meters, but the reader tends to be less effective in the presence of obstacles, liquids, and metals. Fortunately, there are many ways to address the problem. In tagging soda cans, for example, it is possible to tag not the can directly but the cardboard or plastic container which holds the cans. In general, a separation between the antenna and the surface can greatly improve performance. Another approach that has received attention recently is the use of the near field in UHF. In other words, by designing the tag antenna to have an additional "coil element" to pick up the magnetic field, it is possible in principle to energize the tag from a UHF reader even in the presence of liquids and metals. However, despite some early successes on this front, there are only a few commercial products available today that exploit this capability. Yet products do exist today, ranging from tags that are optimized for placement on pharmaceutical bottles to tags optimized for cases.

Readers

There has been a great deal of progress in the cost and performance of readers since the inception of EPC GenII. Today, readers exist in several forms: fixed, mobile, and hand-held. There are also credible reports of readers being sold in the market for under $500.00 [12]. These readers support a variety of connection modalities: from Ethernet to wireless. In the mobile and hand-held category, battery life, hitherto a key challenge, seems to have been addressed adequately. These readers are showing good performance for reasonable periods of time between charges.

Most fixed reader platforms are designed for multiple antennas. Fixed readers are increasingly based on the power-over-Ethernet (POE) standard. However, despite this, the installation costs of readers can sometimes be prohibitively high. In some cases, fixed reader installation rivals the cost of the reader itself, especially in locations like warehouses.[1]

Despite advances in hand-held and mobile readers, an important problem today is that these systems do not support any location technology. Without location, it is hard to interpret the meaning of a reading event. For example, a hand-held reader that reads a tag in a truck at the dock door is reporting a very different action than is a reader that reads a tag on the shop floor. The context is clearly related to location, which is today not recorded.

---

[1] Personal confidential conversation with an industry expert in 2006.

System software

Read events that emanate from readers must be processed, interpreted, and reacted to. Here too there are today commercial products that provide a great deal of functionality. Generally there are layers of activity. The first layer consists of the management of the device. EPC GenII devices have a number of configuration parameters and control actions that must be executed for optimal operation. These include device discovery, the setup of such parameters as the session of a reader, and whether it is in dense mode, and operational monitoring to ensure that the device is operational. For the most part, these functions have thus far been provided with proprietary commands. Recently, the LLRP standard within EPCglobal has offered hope in the basic operation of devices. In addition, other groups like Reader Management and Discovery, Configuration & Initialization (DCI) are beginning to standardize discovery and management.

The second layer is data acquisition from the RFID reader, an area where several factors conspire to make RFID unique and different from other sources of data. They include false positive reads (though greatly reduced in EPC GenII compared with Gen 1), false negative reads, filtering based on sub-parts of the electronic product code or other ID data in the tag, and duplicate suppression. In addition, within the realm of data acquisition is the key idea of password distribution. In EPC GenII, it is possible, upon seeing a particular tag that happens, for example, to have memory, to access that memory within the basic inventory process. For this, the reader must have the necessary password for that tag at its "fingertips." LLRP provides the means to arm the reader with such functionality proactively. The second layer of software must also accomplish this important goal. Some parts of this functionality are today addressed by EPCglobal's LLRP, and some by ALE.

The third layer of software deals with actions. While RFID data by itself is interesting, the data itself has a "half-life;" the value of the data diminishes very rapidly with time. For example, a truck that is leaving with the wrong pallet must be warned before it leaves; recording the event and noting it in the company's collective cognition afterwards is a pyrrhic action. For such actions to be performed, it is necessary that workflows be executed in real time. It is tempting to map these functions to existing workflow engines, rules engines or complex-event-processing engines. However, RFID data poses unique challenges that existing systems might not be able to handle at a native level. Missed reads (say from a broken tag or reader), stray reads (tags that are validly picked up, say, by a reader at a neighboring dock door), and the need to fuse different data sources (for example, a motion sensor might indicate that nothing really came through a dock door, and the tags seen by a reader were simply a stray pallet passing nearby) make RFID data especially "corrosive" to existing systems. RFID data fundamentally requires a great deal of "data healing" and inferencing in order for it to be made valid [13].

The fourth, and highest, layer of software deals with interfacing. Here, EPCglobal has developed a powerful standard called the EPC Information Service. EPCIS enables companies to interface with each other and share data about goods tagged with EPC tags. EPCIS can also be used to connect different systems within the company – say an analytics engine – with the EPC data.

Network software

Companies also need to interact with each other. It is safe to assume that two large companies, like Wal-Mart and P&G, do know each other, and can do so through EPCIS. Issues like security and authentication can also be addressed through predetermination between known trading partners.

However, there are situations in which companies might not know each other. Consider, for example, that a construction company has purchased a lot of commodity wood from a wood dealer. The company then seeks some additional information about the wood, which the wholesaler does not provide. The wood must be traced to the original grower. By reading the RFID tags on the wood, the purchasing company can extract an EPC. The Object Name Service (ONS) developed by the Auto-ID Center and EPCglobal in 2005 provides a mapping from the EPC to the original grower, enabling a direct connection when necessary.

The ONS, however, covers only some very basic needs of a larger set of questions that are collectively referred to as "discovery." Today, the general discovery problem remains surprisingly difficult to state, let alone solve. In recent months, the Architecture Committee of EPCglobal has developed a framework for describing the many permutations of the general discovery problem. EPCglobal is currently initiating the standards development process for solving the more tractable versions of this problem.

2.2   On the future of RFID technology

While it is impossible to truly predict the future direction of a technology as fundamental and vital as RFID, it is possible to identify gaps where innovation will have immediate benefits. We do so below.

Innovations in RFID tags

Despite the significant advancement in the EPCglobal GenII protocol, there are several issues that bear closer attention.

Security
The first issue is around security. EPC GenII does not support cryptography, and few existing products have cryptographic functions. Today, all access control is executed using passwords. Passwords, however, are difficult to manage. Passwords that have been released to the public cannot be revoked, and, since passwords must by definition be passed on, the security of a password-based system cannot be

guaranteed. Fortunately, this is not a serious problem with consumer goods like shampoo. However, as RFID tags become widespread in applications like airline engines, a more robust security scheme will become necessary. Weis *et al.* [14] describe security measures and constraints with conventional passive RFID.

There has been a significant amount of research in recent years on cryptographic approaches for RFID. The trade-offs in passive RFID are severe: chips must not consume too much power because they will not perform well; they must not be too large because they will be expensive; and the computation must not take too long because RFID communication tends to be unreliable. Significantly, Feldhofer *et al.* [15] and Poschman *et al.* [16] have redesigned the Advanced Encryption Standard (AES) and Digital Encryption Standard (DES), respectively, to fit within a few thousand gates. The timing and implementation issues of these approaches will need verification (for example, some of these implementations demand interweaving of challenges and responses to overcome the long computation time) but they are key steps towards advanced security in RFID. Using such features, it will be possible to achieve higher levels of security in RFID. New RFID tags might include features like the following.

- **Tag authentication.** It will be possible for a reader to verify whether a tag is truly the one it claims to be. This will provide the rapid confirmation which will truly assist in the fight against counterfeit products.
- **Reader authentication.** The tag will be able to verify that a reader is truly the one it claims to be. With this functionality, tags might be able to avoid the problem of being promiscuous; they will respond with their ID only to those readers which they deem qualified to know their IDs. This will assist in protecting privacy.
- **Session-based security.** Foley describes an approach to security in which there are no passwords to be lost, and in which the tag is "open" for commands for only a limited period. This will further protect tags from illegal operations.

There is a great deal of scope for the improvement of RFID systems in the areas of privacy. Ideas will range from the subtle, like the blocker tag concept [27], to full-blown cryptography as I have described it above.

Protocol extensions

As mentioned earlier, passive RFID systems have limitations in terms of their command sets. In addition to security, another area deserving of attention is the continuum from passive RFID tags, via semi-passive tags, to active tags. In [18] we describe a vision whereby passive RFID tags, semi-passive tags, and active RFID tags would be part of one continuum. While there has been progress in standardization of semi-passive RFID within EPCglobal, an intellectually sound strategy for including active tags within the same continuum is still needed.

Figure 2.2 shows some of the different, incompatible protocols which could be used as active tags. The question remains that of how they would be compatible with passive and semi-passive RFID tags.

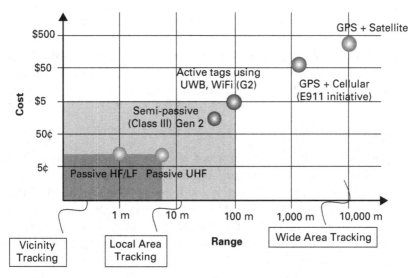

Fig. 2.2   The range of identification technologies.

Electromagnetic performance of RFID

In the presence of liquids and metals, the performance of UHF RFID tags diminishes considerably. There is a great deal of innovation yet to be explored in the performance of RFID tags. Already, the concept of near-field UHF has gained some success [19]. The use of meta-materials that use internal material structure to create favorable antenna properties is another approach that offers some hope [20]. Other similar innovations ranging from antenna design to packaging solutions remain of interest.

Manufacturing

Today, RFID tags are applied after manufacturing, usually in the DC before shipping to retailers. This makes tags expensive for a number of reasons. First, the tags themselves need to be converted to labels with adhesives, anti-static coatings, and a printable area. In the future, tags will likely be inserted into packaging materials. It will also be necessary to ascertain whether tags can be incorporated into blow-molding, injection-molding, and other manufacturing operations.

Sensors

Despite the promise of RFID-based sensors, the technology has not materialized commercially, especially in the passive domain (temperature sensors on battery-assisted tags do exist). There are several challenges in the use of sensors in tags. The first, of course, is power – transduction usually consumes significant power. There is a great need for new techniques to scavenge power for RFID [21]. The second is memory. Data from sensor logs can be copious and needs to be stored. One approach would be to develop an engine that "integrates" the temperature log and reduces it to a single number which represents the "Arrhenius equivalent" of a temperature sequence. In other words, the string of temperatures could be compressed to a single representative number that captures the kinetics of the chemistry.[2] An alternative is to use traditional compression. The third problem is one of uploading the temperature log. RFID has limited bandwidth, and it might take a long time to upload lots of data. Much research waits to be done in these areas.

Reader innovations

There are several fronts along which research and development in RFID could contribute to the RFID industry and its applications.

RFID readers today are generic and follow a "one size fits all" paradigm. This is disadvantageous because readers are overloaded with unnecessary functionality to serve unknown functions. The cost–benefit analysis of the use of RFID depends greatly on the cost of readers. There is an emerging need for a larger class of specialized readers, which we refer to as *application-specific readers* (ASRs). Several classes of ASRs are possible.

1. Lightweight WAN-enabled portal readers. Surprisingly, the cost of wiring RFID readers and connecting them to Ethernet is a significant part of RFID deployments. There is a need for low-power RFID readers that are, for example, battery-powered, motion-triggered, and equipped with a 2.5G or 3G backhaul system. The installation of such a system would be extremely inexpensive. The research question in the design of a system of this type would be related to power consumption and performance versus cost.
2. Location-aware mobile readers. Mobile RFID readers will find great use in RFID systems. Already, several companies are offering mobile reader systems. However, it is hard to interpret the read events from a mobile reader. A read event in a backroom has a different context than a read event in a truck. There exists a need to, on the one hand, make the reader location-aware using, for example, a local positioning system like a WiFi positioning System (WPS). On the other hand, there is also a need for a local inference system that interprets this data and provides the logical hooks to make the reads useful.
3. Unmanned vehicle readers. At the more ambitious end of the spectrum, there might exist a need for readers that are on unmanned vehicles like robots. This might occur, for example, in the cataloging of a yard full of containers or in finding library books [22].

Software

There is great scope for research in software systems for RFID. Several areas present themselves, but I will discuss one in particular. As discussed in the section on system software, false positives and false negatives create a challenge in RFID

---

[2] Conversation with David Brock, Ph.D., Laboratory for Manufacturing Productivity, MIT (2006).

at all levels of the software. One area where this can be particularly important is in inventory systems. Consider a door reader, at the transition between the back room and the sales floor, which has a 90% read rate. Simply using this data in an inventory system can lead to a 10% overestimate of the inventory in the back room because some exiting product will not be detected. However, by considering the entire ensemble of data, such errors can be detected and corrected fairly extensively. For example, let us say that tags are also detected at check-out. If a tag that was missed at the door is seen at check-out, then it is possible to update the back-room inventory appropriately. The fact that EPC tags have unique IDs makes this form of individual inferencing both likely and feasible. Yet there are no good methods for capturing such logic today in a DB system or an application. I refer to this as *inferencing* or *data-healing*, and expect it to be a core part of RFID systems in the future.

2.3   Applications

The applications of RFID cannot truly be known today – so fundamental is the act of automatically identifying an item in the physical world. The majority of applications in the first wave of RFID have been in the supply chain, so I start by summarizing applications there. Before doing so, however, a few words on the distribution of physical assets in a system.

The movements of "things" can be viewed in a three-dimensional space as shown in Fig. 2.3. The two dimensions that each item in the supply chain traverses are time and space. The TAG EPC axis in Fig. 2.3 represents the fates of two individual items. Bundles of these trajectories are what we refer to as the supply chain. We can now define certain terms that are common in supply chain parlance. The word "inventory" refers to a horizontal slice through this space, and answers the question *"what is at location X now?"* This is shown in Fig. 2.4. An RFID reader, meanwhile, takes an "inventory snapshot" in its field of view, which may

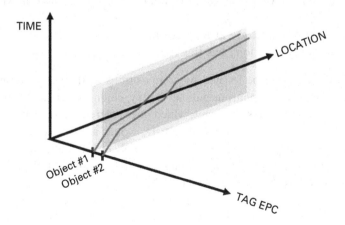

Fig. 2.3   The supply chain can be seen as a 3D space.

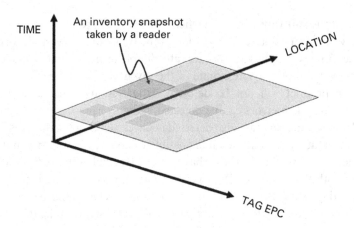

Fig. 2.4   An inventory slice.

be a small corner of a DC. This is shown as a rectangular patch in Fig. 2.4. By tiling together several RFID snapshots, it is possible to infer the inventory of a location or a set of locations in a system. By further taking many such inventory slices, it is possible to reconstruct the trajectories of all known objects in the space, much like MRI can construct a map of the brain from many slices. In this sense, RFID can be seen as a form of real-time MRI of the brain.

Just about every application of RFID that I have come across can be viewed at some level as a querying of this corpus of data. The questions can be along the lines of

- Finding: Where is a certain item now?
- Tracking: Tell me where this item goes.
- Tracing: Tell me where this item has been.
- Positive assurance: Assure me that this item has always been within the following "legal locations."
- Negative assurance: Assure me that this item has never left the following legal locations.
- Counts: How many items are at this location now?
- Time-intersections: How long has this item spent at this location?

Complex questions can be formed by compositions of these simple ones. For example, someone in the pharmaceutical supply chain might ask "Of all the objects I have received in this shipment, which ones entered the supply chain from locations unknown?" This question, obviously, would address concerns around counterfeiting. Similarly, someone in the retail supply chain might request "Please let me know if any item leaves one station in the supply chain but fails to appear in the next." This is illustrated visually in Fig. 2.5.

In this manner, applications can be viewed as recognizing patterns in the time–space–ID data corpus which RFID delivers, and then, let's not forget, acting on it as shown in Fig. 2.6.

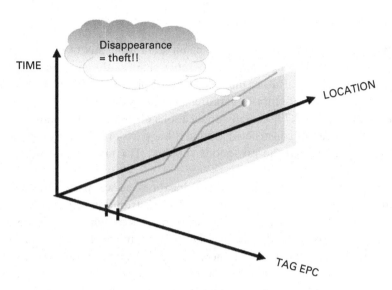

Fig. 2.5   Sudden disappearance implies theft.

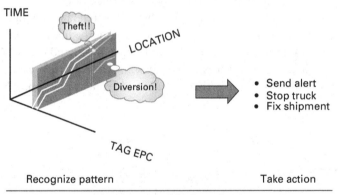

Fig. 2.6   RFID applications.

It is the action, which in some ways has little to do with RFID *per se*, which appears to be the most daunting impediment to the adoption of RFID. This is simply because new actions imply new business processes, and business-process change has long been an area where industries have struggled [23], especially in the context of RFID [24]. For this reason, in adopting RFID, the ideal approach is often to target applications with the most dovetailed, least disruptive business-process change. This runs counter to the view that the applications targeted must be those with the greatest ROI. In many ways, this may be an illustration of the innovator's dilemma: disruption often needs its own sandbox in which to blossom, unfettered by the constraints which are the legacy of ongoing concerns [25].

Applications in the supply chain

Applications of RFID in the supply chain seem to mostly fit the model we have described above. In some ways, the supply chain is the most mundane area of application of RFID. Rather than list all the applications of the supply chain, let's use a matrix in the attempt to capture some of its diversity.

The table in Fig. 2.7 lists many applications of RFID in the retail supply chain. Similar tables can be constructed around other parts of the supply chain – be it the manufacturer or even the raw-goods supplier. The table is laid out roughly as follows.

The columns represent the different locations as we march down the supply chain. The terms are defined as follows. DC: distribution center; BR: back room of the store; SF: sales floor of the store; Storage: areas where goods are stored for layaway, for example. Terms like DC–BR represent the transport of goods between those two entities. In essence, the supply chain can be viewed as a series of storage operations, assembly/disassembly operations (assembling goods into a pallet, disassembling cases from a pallet, for example), and transport operations.

The rows represent different concerns in the supply chain. Time concerns are prevalent with perishable goods and in terms of maximizing turns. Quantity concerns are paramount in terms of having enough inventory to meet demand. A shortage manifests itself as a stock-out. Configuration refers to having not just the right goods, but the right combination of goods along the lines of various supply chain rules. For example, it is preferable not to have foods and chemicals shipped on the same pallet, and it is necessary to lock fire-arms in secure locations in a store. Regulation refers to rules that are derived from regulation. They usually

	DC	DC–BR	BR	BR–SF	SF	Storage
Time	• Dwell time • Code date	• Timeliness	• Dwell time • Code date	• Code date (seasonals, expity date promotions)	• Customer experience	• Locating goods like special orders rapidly
Quantity (Inventory)	• Automatic receiving • Location • Replenishment	• Shrinkage	• Assured receipt • DSD • Location • Replenishment	• In-flow measurement • Shrinkage	• Automatic replenishment	• Managing yards where bulk goods are stored
Configuration	• Code/rule compliance (food, chemicals, firearms, etc.)	• Code/rule compliance	• Code/rule compliance	• Display/ collateral	• Ensemble availability	• Finding peripherals for electronics
		Re-usable assets				
Regulation		Pedigree Code date/age			• Age check	• Ensuring that returned goods like drugs don't expire
Sensors		Temperature/freshness				
		Shock				

Counterfeit, parallel trade, etc. not captured

Fig. 2.7   A map of the retail supply chain.

apply to food items and to pharmaceutical items. Finally, there is a growing interest in the use of sensors like temperature sensors and pressure sensors to monitor supply chain performance and ensure advanced condition-based monitoring.

Many of the key applications in the supply chain as illustrated in the chapters that follow can be mapped onto a tableau of this form. For example Hardgrave (Ch. 9) refers to end-to-end supply chain visibility and applications like promotional execution and claims management [26] [27] that are also proving to be increasingly promising. McFarlane *et al.* describe how the Aero-ID programme at Cambridge University addresses issues of configuration, regulation, and sensors specific to the aero-defense industry. By contrast Emond (Ch. 11) refers to the need in "cold chains" to respond to FDA requirements under SC31 that require "real-time" temperature tracking, which are challenging to integrate from multiple sensors across a supply chain.

Beyond the supply chain

The applications of RFID beyond the supply chain are clearly impossible to capture in a brief section. In principle, any action that involves the movement of goods or people might benefit from RFID if ethically and suitably deployed. I list some examples below.

Asset tracking

A vast class of applications in RFID is related to assets – equipment, peripherals, trucks, compressors, containers, etc. – in a variety of arenas including factories, yards, institutions (like universities and hospitals), facilities (like oil-rigs, maintenance depots, and ports) and the field (like battle fields and construction sites.) Many of these areas are metal-heavy or need long-range RFID, which is often beyond the scope of passive RFID. Semi-passive tags and active tags have and will certainly provide the necessary tracking capability for many of these applications, but a glut of standards, combined with the lack of a clear continuum between passive tagging and powered tagging, remain obstacles in adoption.

Battling counterfeits

As Fleisch *et al.* report in Ch. 12 (see also [28]), there are myriad application arenas for RFID in battling the scourge of product dilution and counterfeits. Passive RFID with the current functionality of the EPC GenII tag primarily endows a product with a history. This provides an indirect way to infer whether a product has remained within the legitimate supply chain, and to assure its pedigree. The same technique is also useful in detecting product diversion or parallel trade. In general, the approach today might be to detect these problems at a macroscopic level rather than pinpointing the problem. The current state-of-the-art in RFID does not yet permit on-the-spot verification of whether a product is valid because cryptography on passive tags has not yet advanced to the level where it is economical or feasible on low-cost RFID tags.

Tracking in hospitals

The hospital is an institution where safety, cost, and regulations intertwine at an extraordinarily heightened level of awareness. Keeping track of patients, equipment, and pharmaceutical products is a very important function. Applications include tracking infirm patients or babies, tracking equipment like stretchers and ECG machines, tracking instruments in operating rooms, tracking the dispensation of medicines and implants like stents, tracking documents like X-ray charts, ensuring cold-chain compliance, tracking blood and tissue, and tracking lab samples.

Baggage tracking

While lost baggage in airlines remains both an economic and a logistical irritant both for airlines and for passengers, new security concerns take the tracking of baggage to an entirely new level of importance. Today, several successful pilots have demonstrated that baggage tracked with passive RFID tags does achieve acceptable read rates and the ROI is indeed positive [29]. However, many deployments remain pilots [30].

2.4   Conclusions

I have argued in this chapter that, though the RFID industry can take pride in its achievements over the last decade, there is much that remains to be done. It is the duty of technologists to enable applications, and the applications domain in RFID is very much in its infancy. So, in fact, is the technology itself, for, though a few components have seen giant strides recently, the system has not advanced as a whole to its fullest possible extent. I have argued that this will occur in waves, and that we are in the first wave. The future of RFID is exciting, and there are great opportunities for the creative and the entrepreneurial.

2.5   References

[1] **Sarma, S.**, "A History of the EPC," in *RFID: Applications, Security, and Privacy*, ed. S. Garfinkel and B. Rosenberg, pp. 37–55 (Addison-Wesley Professional, New York, 2005).

[2] **EPCglobal Inc.**, *The EPCglobal Architecture Framework*, 2005-07-01. Note: EPCglobal standards are available on-line (http://www.epcglobalinc.org/standards).

[3] **EPCglobal Inc.**, *EPC Radio-Frequency Identity Protocols Class-1 Generation-2 UHF RFID Protocol for Communications at 860 MHz–960 MHz Version 1.0.9* (2005).

[4] **EPCglobal Inc.**, *Low Level Reader Protocol (LLRP) Version 1.0* (2007).

[5] **EPCglobal Inc.**, *The Application Level Events (ALE) Specification Version 1.0* (2005).

[6] **EPCglobal Inc.**, *EPC Information Services (EPCIS) Version 1.0 Specification* (2007).

[7] **Sarma, S. E.**, *Towards the Five-Cent Tag* (Auto-ID Center, Cambridge, MA, 2001) (http://www.autoidlabs.org/uploads/media/mit-autoid-wh-006.pdf).

[8] **Sarma, S. E.**, and **Swamy, G.**, *Manufacturing Cost Simulations for Low Cost RFID Systems* (Auto-ID Center, Cambridge, MA, 2003) (http://www.autoidlabs.org/uploads/media/MIT-AUTOID-WH017.pdf)

[9] **Roberti, M.**, "SmartCode offers 5-Cent EPC Tags," *RFID Journal* (2006).

[10] **Lowe, F.**, **Craig, G. S. W.**, **Hadley, M. A.**, and **Smith, J. S.**, *Methods and Apparatus for Fluid Self Assembly*, US patent 7172789 (2007).

[11] **Syed, A.**, **Demarest, K.**, and **Deavours, D. D.**, "Effects of Antenna Material on the Performance of UHF RFID Tags," *IEEE RFID 2007*, March 26–28, 2007 (Grapevine, TX).

[12] **Alien Technology Corporation** *Registration Statement* (Securities and Exchange Commission, Washington, DC, May 31, 2006) (http://ipo.nasdaq.com/edgar_conv_html%5C2006%5C05%5C31%5C0001193125-06-121612.html).

[13] **Sarma, S. E.**, "Integrating RFID," *ACM Queue*, 2(7):50–57 (2004).

[14] **Sarma, S. E.**, **Weis, S. A.**, and **Engels, D. W.**, "RFID Systems and Security and Privacy Implications," in *Proceedings of Workshop on Cryptographic Hardware and Embedded Systems*, pp. 454–470 (Springer, New York, 2002).

[15] **Feldhofer, M.**, **Dominikus, S.**, and **Wolkerstorfer, J.**, "Strong Authentication for RFID Systems using the AES Algorithm," in *Proceedings of Workshop on Cryptographic Hardware and Embedded Systems*, Boston, USA, August 11–13, 2004, pp. 357–370 (Springer, New York, 2004).

[16] **Poschmann, A.**, **Leander, G.**, **Schramm, K.**, and **Paar, C.**, "DESL: An Efficient Block Cipher for Lightweight Cryptosystems," in *Workshop on RFID Security 2006, RFIDSEC '06*, ed. S. Dominikus (Graz, July 2006) (http://events.iaik.tugraz.at/RFIDSec06).

[17] **Juels, A.**, **Rivest, R. L.**, and **Szydlo, M.**, "The Blocker Tag: Selective Blocking of RFID Tags for Consumer Privacy," in *8th ACM Conference on Computer and Communications Security*, ed. V. Atluri, pp. 103–111 (ACM Press, New York, 2003).

[18] **Sarma, S.**, and **Engels, D. W.**, *Standardization Requirements within the RFID Class Structure Framework* (Auto-ID Center, Cambridge, MA, 2005) (http://www.autoidlabs.org/uploads/media/AUTOIDLABS-WP-SWNET-011.pdf).

[19] **Nikitin, P. V.**, **Rao, K. V. S.**, and **Lazar, S.**, "An Overview of Near Field UHF RFID," *IEEE International Conference on RFID*, pp. 167–174 (2007) (http://www.ee.washington.edu/faculty/nikitin_pavel/papers/RFID_2007.pdf).

[20] **Dacuna, J.**, and **Pous, R.**, "Near-field UHF Tags Based on Metamaterials Concepts," *EU RFID Forum/RFID Academic Convocation* (2007) (http://www.rfidconvocation.eu/Papers%20presented/Technical/Near-field%20UHF%20tags%20based%20on%20metamaterials%20concepts.pdf).

[21] **ISA**, "Power Scavenging Sensors Shun Batteries" (2006) (http://www.isa.org/InTech Template.cfm?Section=Technology_Update1&template=/ContentManagement/Content Display.cfm&ContentID=54734).

[22] **Ehrenberg, I.**, **Florkemeier, C.**, and **Sarma, S.**, "Inventory Management with an RFID Equipped Mobile Robot," *Proceedings IEEE Case 2007* (Scottsdale, AZ, 2007).

[23] **Brynjolfsson, E.**, **Austin Renshaw, A.**, and **van Alstyne, M.**, *The Matrix of Change: A Tool for Business Process Reengineering* (MIT Press, Cambridge, MA, 1996) (http://ccs.mit.edu/papers/CCSWP189/ccswp189.html).

[24] **Bamforth, R.** (IT Analysis Communications Ltd.), "Business Processes, not Technology, Show Who's BOSS in RFID" (2004) (http://www.it-director.com/technology/mobile/content.php?cid=7441).

[25] **Christensen, C.M.**, *The Innovator's Dilemma: When New Technologies Cause Great Firms to Fail* (Harvard Business School Press, Cambridge, MA, 1997).

[26] **Subirana, B., Sarma, S., Ferguson, C., Langford, S., Spears, M., Jastremski, G., Dubash, J.,** and **Lee, R.**, *EPC Changing the CPG Industry: Improving Retail Promotional Execution* (EPC Global, Princeton, NJ, 2006) (http://autoid.mit.edu/documents/Promotional%20Execution%20Vignette%20 ApprovedV3.pdf).

[27] **Subirana, B., Sarma, S., Ferguson, C., Supple, J., Spears, M., Lee, R., Dubash, J., Mason, M., Langford, S.,** and **Roth, L.**, *EPC Changing the CPG Industry: Electronic Proof of Delivery* (EPC Global, Princeton, NJ, 2006) (http://autoid.mit.edu/documents/EPOD%20Vignette%20ApprovedV2.pdf).

[28] **Staake, T.R.**, *Counterfeit Trade – Economics and Countermeasures*, Ph.D. thesis No. 3362, University of St. Gallen (2007).

[29] **IATA**, *RFID Trials for Baggage Tagging* (IATA, Geneva, 2006) (http://www.iata.org/NR/rdonlyres/D319ADC0-ED5D-447E-9EEB-6CA1179C6BD9/0/RFIDtrialsfor baggagetagging.pdf).

[30] "Years Before RFID Baggage Tracking Takes Off, RFID Update," *The RFID Industry Daily* (November 6, 2006).

3 RFID tag performance optimization: a chip perspective

Hao Min

From an historical perspective it is the small CMOS integrated circuits (versus inductively coupled transponders) which incorporated microwave Schottky diodes that made it possible to manufacture small passive RFID tags. In this chapter it is demonstrated how recent improvements in CMOS technology (0.18u and later) make it possible to use an inexpensive MOS transistor for the EPC GenII/ISO 18000-6c compliant transponders that operate in the UHF band. The basic variables for designing low-power (high-readability) RFID tags and details of incorporating temperature sensors into semiconductor chip designs at the Auto-ID Labs, Fudan University, Shanghai are described.

This chapter explores chip design principles that affect the performance of RFID tags. The metrics of tag performance will be illustrated and the corresponding optimization technologies will be introduced. Many early implementation issues for RFID tags involved low read rates such that tags did not "wake up" in response to interrogator signals and transmit their IDs, especially with transponders operating at UHF frequencies that are subject to interference from liquids and metals.

3.1 Metrics of tag performance

In RFID applications, tag performance will directly influence the success of the whole system. Understanding the metrics of tag performance is important in order to foresee the overall system performance. This section will analyze the metrics of tag performance, including read range, read rate, communication speed, etc.

In a free-space propagation environment, the power received by an RFID tag antenna can be calculated using the Friis free-space propagation equation as

$$P_{\text{tag}} = \text{EIRP} \times G_{\text{tag}} \left(\frac{\lambda}{4\pi r} \right)^2 \tag{3.1}$$

RFID Technology and Applications, eds. Stephen B. Miles, Sanjay E. Sharma, and John R. Williams. Published by Cambridge University Press. © Cambridge University Press 2008.

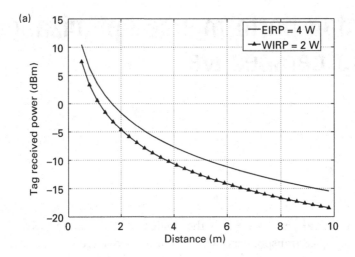

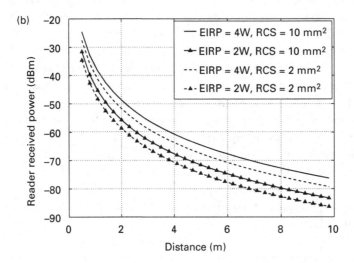

Fig. 3.1   (a) Read range versus tag received power. (b) Read range versus tag backscattered power.

where P_{tag} is the power received by the tag, G_{tag} is the gain of the tag antenna, EIRP is the power transmitted by the reader, λ is the electromagnetic wavelength, and r is the read range.

Read range

One of the most important characteristics of an RFID tag is its read range. One limitation is the maximum read range at which the tag receives just enough power to turn on and scatter back. The maximum distances for various powers received by the tag are given in Fig. 3.1(a).

According to Eq. (3.2), the maximum operating distance can be calculated from

$$r = \frac{\lambda}{4\pi} \sqrt{\frac{\mathrm{EIRP} \times G_{\mathrm{tag}}(1 - |\Gamma|^2)}{P_{\mathrm{th}}}} \qquad (3.2)$$

where P_{th} is the minimum threshold received power which just powers the tag and Γ is the reflection coefficient.

Another limitation is the maximum read range at which a reader can just detect the scattered signal. It is determined by reader sensitivity and backscattered energy. Figure 3.1(b) shows the maximum operating distances for various backscattered powers. Greater operating distance needs higher reader sensitivity. In general, reader sensitivity is high enough that the read range is determined by the threshold received power.

Singulating speed

The singulating speed represents the number of tags being singulated in one second in RFID systems. It is determined by many parameters, such as downlink speed, uplink speed, channel bit error rate (BER), the anti-collision algorithm, and the total number of tags being powered.

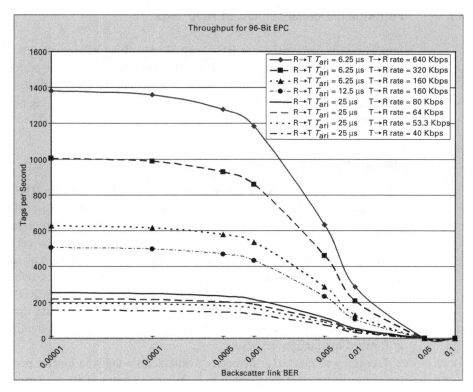

Fig. 3.2 Singulated tags per second versus backscatter link BER.

For different T→R (tag to reader) data rates, the numbers of tags singulated per second have almost an order of magnitude difference. There are about 1,580 tags being read when the data rate is 640 kbps, but only about 200 tags at 40 kbps in the non-error-condition mode. The singulating speed varies with the backscatter link bit BER (Fig. 3.2). So, increasing the data rate and decreasing the BER in any feasible ways are direct methods by which to improve the singulating speed.

Read rate

The read rate is defined as the percentage of attempts on which a reader can successfully access tags. This is another problem suffered in RFID technologies. RFID readers are not always able to read chips on a 100% basis. One of the primary reasons is poor positioning, sometimes in an area called a "null spot," which results in a lack of sufficient energy. This phenomenon is like "Swiss cheese" insofar as there are bubbles where the RF field is not strong enough to operate the tag. Other reasons have to do with interference from metals, water, other RFID tags, or anything that generates electromagnetic energy.

3.2   Performance enhancement of RFID tags

Improving the performance of RFID tags is very difficult. Many factors should be considered and optimized. Different approaches to optimizing the performance of tags will be introduced in this section. Power management, anti-collision algorithms, and antenna and chip matching are important factors for tag performance optimization.

Chip power modeling

Figure 3.3 (below) shows the block diagram of an RFID tag chip. The antenna is the only external component of the transponder. The tag chip consists of four main blocks: an RF front end, an analog front end, a base-band processor, and an EEPROM section. The RF front end is responsible for power recovery, demodulation of the incoming RF signal, and the backscatter transmission of return data. The analog part is to generate the system clock, current/source reference, and power-on-reset (POR) signal, etc. The base-band section and EEPROM memory handle power management, data recovery, operating protocols, and user-available data storage. In the RFID tag system, the signal flow can be divided into two paths of forward link and reverse link.

According to Eq. (3.1), the power available to a tag is in inverse proportion to the square of the operating distance. In order to maximize the read range of a tag, the power consumption of the tag should be lowered as much as possible.

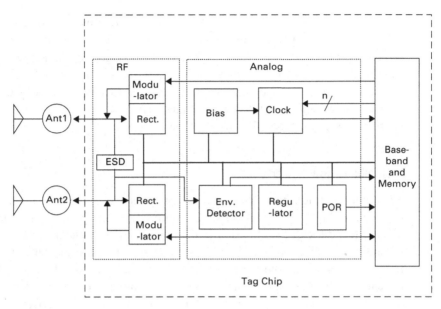

Fig. 3.3 RFID tag chip architecture.

Energy conversion efficiency

A rectifier is used to generate the power supply for the whole chip from the incoming RF signal, which is usually implemented by a voltage multiplier. The energy conversion efficiency of the rectifier is another limitation on the tag's read range. It is about 20% in common rectifier structures, but the conversion efficiency can be up to 50% when dynamic substrate and threshold voltage compensation techniques are employed. The received power of a rectifier is influenced by the matching network and operating distance. The output energy from a rectifier is decided by its efficiency, stage number, operating bandwidth, etc. So the optimization of this kind of multiplier must take into account many factors. These factors interact with each other.

The output voltage and efficiency are calculated as follows:

$$V_{\text{out}} = -V_{\text{j,dc}} \frac{R_{\text{L}}}{R_{\text{S}} + R_{\text{L}}/(2N)} \tag{3.3}$$

$$\eta = \frac{\text{dc output power}}{\text{incident RF power} - \text{reflected RF power}} \tag{3.4}$$

Power management

Passive tags can communicate to the reader without a battery by using power received from the tag antenna. The power received can diminish to a few tens of microwatts at a long operating distance. In view of this condition, low-power techniques are absolutely required for passive tag design. In addition, reasonable power management should be considered. For example, in the process of back-scatter signaling, the demodulator, decoder, and any other modules can be shut off, and vice versa. An on-chip capacitor of large capacitance is used to store received energy, and its optimized capacitance can be derived according to priorities for stored energy, process flow or chip size.

Low-voltage, low-power circuits

The ability to read a tag consistently is closely related to the power consumption of the microchip in the tag. A chip that requires more energy will be read less often because it is difficult to get energy to the tag when the waves must travel through materials or when there is interference in the environment. Ultra-low-power techniques, sub-threshold circuits, adiabatic circuits, etc., should be developed in the chip design, especially for the RF/analog front end. This can reduce the tag's power consumption by a factor of 2–10. The UHF RF/analog front end is an interface between the antenna and the base-band processor. The main blocks are listed in the following section.

Demodulator

The demodulator is one of the most important circuit blocks in the forward link. It is used to extract the data symbols which are modulated in the carrier waveforms. In RFID systems, the modulation type from reader to tag is DSB-ASK, or SSB-ASK, or PR-ASK in the EPC GenII protocol. The signal is always shown in the form of amplitude change, so the demodulator is actually an edge detector.

The structure of a simple demodulator is shown in Fig. 3.4. The envelop detector can share the circuit with the rectifier used to generate the power supply for the tag chip. The low-pass filter is used to filter out the carrier ripple noise residue and takes the difference from the two inputs of the hysteresis comparator. Then the received signal is detected and digitized to output.

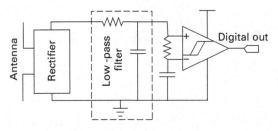

Fig. 3.4   A simple demodulator structure.

The RFID tag components identified below perform the functions specified.

- The clock generator generates the system clock by use of oscillators of some kind.
- The voltage regulator maintains the power supply at a certain level and at the same time prevents the circuit from breaking under large RF input power. As we know, the EM field strength may vary in magnitude by factors of tens or even hundreds at different physical locations [1]. Following the variation of operating distance, the energy received from the tag antenna and the power supply rectified by the voltage multiplier are different. A voltage regulator can regulate the power supply to a relatively stable voltage and keep other analog circuits operating at better performance. A shunt regulator and a serial regulator are always used in tag chip design.
- Power on reset is used to generate the chip power on reset (POR) signal.
- A voltage reference is used to generate some voltage or current reference for the use of front-end and other circuit blocks, such as EEPROM, usually in terms of the reference band gap.

Increasing the singulation rate

The following steps have been identified as means to maximize the number of tags that can be read.

Adjusting the communication speed

Communication speeds can be evaluated by completing a series of command cycles. Let's take a single query cycle into account in the EPC GenII protocol. That is the total time required from interrogator command transmission to receipt of tag response by the reader (measured at the reader's antenna terminals). The total time is the summation of the following values.

1. T_4, defined in EPC GenII protocol, the minimum time between interrogator commands.
2. The length of the query command.
3. T_1, defined in the EPC GenII protocol, the time from transmission to tag response, measured at the tag's antenna terminals.
4. The transmission length of RN16.
5. T_2, defined in the EPC GenII protocol, the time required if a tag is to demodulate the interrogator signal, measured from the last falling edge of the last bit of the tag response to the first falling edge of the interrogator transmission.

The calculated results are shown in Table 3.1.

Supposing that T_{ari} is 6.25 µs, the maximum ideal singulation rate is about 1,800 tags per second. So improving the communication speed is a good choice as a way to increase the singulation rate.

Table 3.1. A single query cycle

T_{ari} (µs)	T_1 max (µs)	T_2 max (µs)	T_4 max (µs)	Query max (µs)	RN16 max (µs)	Total (µs)
6.25	72.3125	21.09375	37.5	243.75	154.6875	529.3438
12.5	142.625	42.1875	75	487.5	309.375	1056.688
25	283.25	84.375	150	975	550	2042.625

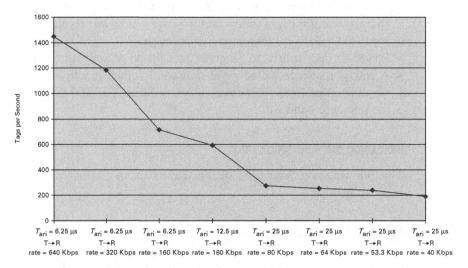

Fig. 3.5   The singulation ratio versus the number of tags.

Choosing anti-collision algorithms

Anti-collision algorithms are used in RFID tags in order to read many tags at a time. When multiple tags are in an interrogatory zone at the same time, the RFID reader needs to distinguish and access those tags which should be read as soon as possible. An effective anti-collision algorithm can reduce the operating time and increase the read rate. Two algorithms, slotted aloha and binary search, are always used in the RFID protocol.

Each anti-collision algorithm has its advantages and disadvantages. The slotted aloha algorithm needs a synchronous signal and a longer time to process when more tags are in collision. Also the discrimination ratio is not as high as that with the binary search algorithm. However, the binary search algorithm has strict requirements for its computing slot and bad security. Figure 3.5 shows the discrimination ratio versus the number of tags in the interrogatory zone.

In the case of RFID technology, anti-collision algorithms need further research, in order to simplify the identification process, increase communication security, and optimize the discrimination ratio.

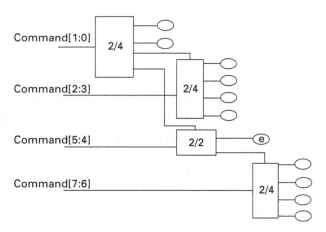

Fig. 3.6 Partial decoder architecture.

Robust decoding methods

In the EPC GenII RFID system, tags shall encode the backscattered data as either FM0 base-band or Miller modulation of a subcarrier at the data rate [2]. According to the command definition, the command is of variable length, from 2 bits to 8 bits, and is transmitted into the tag serially. Robust and quick decoding can increase the tag singulation rate and read rate. A partial decoding method is a good solution. According to the coded format of the command, it can be decoded in 1–3 steps. As Fig. 3.6 shows, an ellipse means finishing decoding a command, and rectangles stand for multiplexers with two inputs. Clearly, in comparison with using a custom decoder with eight inputs, this decoder not only improves the decoding speed, but also lowers the chip area and power consumption.

Improving the read rate

Avoidance of the "Swiss cheese" effect

When RFID systems are applied in an environment with multipath channels (e.g. the indoor case), the signal delivered from a reader may reach a particular point through different paths, but out of phase. If the power levels of these out-of-phase signals are similar, an energy null occurs at this point in space. The positions of these energy nulls are mainly determined by two things: the distribution of multipath channels and the wavelength of the radio signal. The latter is almost fixed (actually different channel frequencies corresponding to different wavelengths, but the discrepancy is quite small because it's a narrowband system) once the frequency band to be used in the application has been chosen. So when the number of multipath channels is large (in the indoor environment, as mentioned before, it's normally hundreds to thousands), an energy null occurs quite often. The appearance of these energy nulls is just like the holes in the famous Swiss cheese.

Moving tags versus quiet-tag strategies

The energy fading caused by the multipath effect is a very serious problem. Tags in the energy nulls cannot get enough activating energy, no matter how near they are to the reader and how long a reading time they are given. A relatively easy way to remedy this situation is to keep tags moving while they need to be read. Experiments have shown that a read rate of above 99% can be achieved in this way. On the other hand, for stationary tags, keeping the reader moving might be a solution.

Power management and state-recovery technology

As mentioned before, a tag receives the energy from the RF signal as soon as it comes within the RF field. So the energy for passive tags is very different from that for active tags, whose energy is supplied by batteries. The instantaneous energy is finite, but the total energy is infinite. So how to reduce the instantaneous power or peak current and make power dissipate evenly over the entire working period [3] is very important. Here the objective of power management is to make the tag chip operate at a low instantaneous power. For base-band processors, the power consumption of each module is described in Fig. 3.7(a) and Fig. 3.7(b) shows the operating situation of each sub-module: from receiving the signal, via executing the command, to back-scattering with timing variety (the dashed line indicates optional activities).

In order to compensate for the influence of the "Swiss cheese" effect, another technology, state recovery, can be utilized in the tag processor. When the power received by the tag is not enough to communicate with the reader, the flip-flops in a processor that has been designed in a particular way can retain the current state and continue to operate until the tag is out of the "energy null." The limitation of this technique is that the time of data retention is dependent on the value of the energy-storage capacitor. So it must be a short period of time.

Antenna and chip matching

The model for antenna and chip matching

In the UHF tag system, the tag antenna and chip impedance can be modeled as shown in Fig. 3.8. The voltage source represents an open-circuit voltage received by the tag antenna. $Z_{ant} = R_{ant} + jX_{ant}$ is the antenna impedance and $Z_{ic} = R_{ic} + jX_{ic}$ is the tag chip impedance. Both Z_{ant} and Z_{ic} are frequency-dependent. In addition, the tag chip impedance Z_{ic} varies with the input chip power. The input power for the tag can reach the maximum value when Z_{ant} and Z_{ic} are conjugated matching. So usually the antenna is designed to match to the chip at the minimum threshold tag power in order to get the maximum read range.

Impedance matching using frequency shifting

Generally, the impedance matching is implemented at the given frequency. For example, the resonance frequency is always the central frequency 910 MHz for the 860 MHz–960 MHz frequency range in the EPC GenII/ISO 18000-6c protocol.

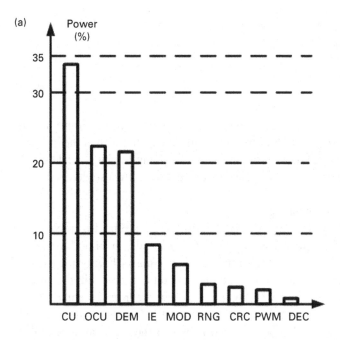

(a)

(b)

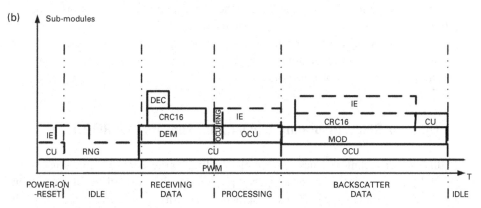

Fig. 3.7 Power management. (a) Sub-module power consumption. (b) Power distribution versus operations procedures.

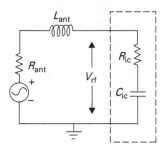

Fig. 3.8 Tag antenna and chip impedance model.

When the carrier frequency is shifted, the impedance matching is destroyed, and part of the energy will be reflected from the tag antenna. The energy received by the antenna P_{ic} can be defined as

$$P_{ic} = P_m g\kappa = P_m g\left(1 - |\Gamma|^2\right) \tag{3.5}$$

where P_m is the maximum energy from the antenna, κ is the power transmission coefficient, and Γ is the voltage reflection coefficient. In addition, when the tags are applied to objects made of metal or containing water, UHF tag antennas are detuned. When the tags are put close to one another, the tag antennas are also detuned.

Dynamic antenna and chip matching

In the case of frequency shifting, the best method would be to employ an impedance that can self-adapt to change. The antenna is designed and attached outside of the tag chip, so a variable-impedance antenna is hardly a possible solution, but a self-adaptive impedance-matching circuit can be realized in a tag chip, which is an impedance-matching network connected from antenna to tag chip. It can adjust its own impedance in order to meet the frequency shifting, and it is less sensitive to detuning by the material to which the tag is attached. The technique can improve read rates on difficult materials and in regions that use different areas of the UHF spectrum for RFID.

In semi-passive or active tags, if the designer doesn't want to save battery power when the reader-to-tag distance is large and never uses the energy from the antenna, this issue is not relevant.

3.3   Sensors for RFID; integrating temperature sensors into RFID tags

In this section, technical issues concerning how to integrate a temperature sensor into an RFID chip will be discussed and a sample design will be described. In general, an RFID is just another type of sensor, namely an identity sensor. In the case of fresh food, or the health industries, temperature tracking linked to identity is crucial. It can protect consumers from products that have suffered breaks in the cold chain, or control the transportation conditions of blood samples [4]. Also acceleration, pressure, humidity, and other sensors may become important for RFID data.

Sensor-enabled tags need to have their own power supply, and fall into the categories of semi-passive or active tags. At the same time, more research needs to be done on RFID systems, such as concerning data transmitting, data processing, new communication protocols, and low-power technology for sensors.

Integrating a temperature sensor into a chip

Temperature sensors provide the detected temperature in either analog or digital format. For analog temperature sensors, the outputs are voltage or current, which change depending on the detected temperature. A digital temperature sensor can be considered as an analog temperature sensor with integrated ADC. It provides digital temperature information. In an analog temperature sensor, the sensor gain is

$$\text{Sensor gain} = \frac{\text{Supply voltage}}{\text{Temperature range}} \qquad (3.6)$$

According to this equation, if the power supply is 1.8 V and the temperature range is −40 °C to + 80 °C, the sensor gain is 15 mV/°C. The lower the supply voltage, the smaller the sensor gain. How to handle small gains and monitor the temperature accurately is the most difficult problem in integrated temperature sensor design. Even for a digital temperature sensor implementing low-power high-accuracy ADC is also a big challenge.

Challenges of temperature sensor in an RFID chip

Our objective has been to review requirements to integrate temperature sensor and RFID tag functionality into a chip. The requirement constraints of RFID tag chip design are equally important for the temperature sensor, such as ultra-low power, small area, low cost, and high performance. At the same time, error cancellation and temperature calibration in a sensor should be implemented at a low cost.

We can summarize the key techniques involved in a temperature sensor according to the following sample digital temperature sensor design steps. The block scheme is shown in Fig. 3.9.

1. The control block is needed to control the power-down mode and the counter. The $\Sigma\Delta$ converters have been proven to be very suitable in low-frequency, high-performance applications.
2. The bit stream at the output is directly proportional to $I_{\text{temp}}/I_{\text{ref}}$; then the signal is led to a counter and the output signal becomes an 8-bit code.
3. Before the signal out of the counter can be read on the monitor as representing temperature, the sensor must be calibrated. The calibration figures are stored in an EEPROM in the other calibration facility; the relation of the 8-bit code and the temperature can be fixed once and for all by means of actual measurement.
4. In the temperature sensor and reference generator block, a PTAT circuit with chopping technique is a good solution to generate the linear function of temperature.
5. Offset cancellation techniques, such as the use of a nested-chopper amplifier, should be embedded within the temperature sensor system.
6. This also reduces the errors caused by the chip's self-heating, so that accurate detected temperatures can be shown.

This chapter has covered the metrics, performance characteristics, and chip layout constraints of integrating sensors on a tag. Starting with this analysis of

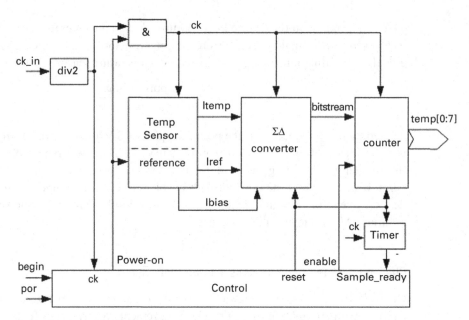

Fig. 3.9   A block schematic diagram of a complete temperature system.

constraints of the RFID tag medium itself provides a basis for considering the characteristics of frequency, protocol, and RF environmental factors that impact RFID systems' performance.

3.4   References

[1] **Zheng Zhu**, *RFID Analog Front End Design Tutorial* (Auto-ID Labs, University of Adelaide, Adelaide, 2005) (http://autoidlabs.eleceng.adelaide.edu.au/Tutorial/RFIDanadesign.pdf).

[2] **EPCglobal**, *Class-1 Generation-2 UHF RFID Protocol, Version 1.0.9* (EPCglobal, Princeton, NJ, 2006) (http://www.epcglobalinc.org/standards/Class_1_Generation_2_UHF_Air_Interface_Protocol_Standard_Version_1.0.9.pdf).

[3] **Abrial, A., Bouvier, J., Senn, P., Rebaudin, M.,** and **Vivet, P.,** "A New Contactless Smartcard IC Using an On-Chip Antenna and an Asynchronous Micro-controller," *Solid-State Circuits, IEEE Journal*, 36(7):1101–1107 (2001) (http://www.autoidlab.fudan.edu.cn/file/published%20paper/A%20low-power%20baseband-processor%20for%20UHF%20RFID%20tag.pdf).

[4] **European Commission Directorate General for Informatics (EC DGIT)**, *Bridge-033546-Annex I* (EC DGIT, Brussels, 2006) (http://www.bridge-project.eu/index.php/workpackage1/en/).

4 Resolution and integration of HF and UHF

Marlin H. Mickle, Leonid Mats, and Peter J. Hawrylak

An act of government, in this case the State of California, is the driving force behind the research topic addressed in this chapter. The use of RFID to establish an ePedigree in the pharmaceutical supply chain brings to a head basic RFID frequency and technology choices that are available from vendors today. This chapter describes how the RFID Center of Excellence at the University of Pittsburgh works with pharmacy distribution and retail as they evaluate different requirements specific to the healthcare life sciences industry.

Different RFID reader environments and the physics of RFID that impact systems performance, including fundamentals of orientation, are characterized in order to explain different findings from pharmaceutical industry HF RFID pilots and from fast-moving consumer goods retail UHF RFID implementations. Alternatives of HF and UHF, namely near-field and far-field RFID options, are explored (the analogy is to compare an RF environment that is like "a prisoner in a cell" with an RF environment like "a bird in the sky"), and performance models are presented and evaluated with respect to constricted orientation and distance in real-world scenarios. A systematic layered approach for analyzing RFID interrogator-to-tag RF protocols is proposed, an insight that, should it be adopted by reader manufacturers, would dramatically improve RFID reader interoperability and testing. Recommendations are made with respect to modeling RFID systems performance and where pilots can help prepare the way for full implementation of RFID systems.

In an EPCglobal HealthCare Life Sciences "bake-off" in the summer of 2006, just when UHF was getting traction in the FMCG industry, US pharmaceutical distributors selected HF as a result of the fact that RFID technology suppliers were able to demonstrate 100% read accuracy in the mixed totes that are typical of drug shipments to individual pharmacies. The use of electromagnetic (EM) analysis and simulation tools including computer-aided-engineering (CAE) tools to evaluate a 13.56-MHz inductor design for an RFID product is described. The CAE software uses Maxwell's equations to analyze planar circuits; experiments are conducted in a rectangular shielded box such as an anechoic chamber. Questions regarding what can realistically be accomplished with such tools, and where specialized expertise is likely to be required in designing RFID packaging for downstream systems efficacy, are addressed.

RFID Technology and Applications, eds. Stephen B. Miles, Sanjay E. Sharma, and John R. Williams. Published by Cambridge University Press. © Cambridge University Press 2008.

4.1   Introduction

The increases in adoption of RFID and the supporting technologies have introduced variations in supply chain objectives that require, at least for the present, choices that are not necessarily synergistic. At the simplest level, the particular choice under consideration is between a basic license-plate scenario with **100% readability under extended conditions** or a *pedigree* that can be documented under conditions of **100% absolute readability**. In the context of this chapter, readability involves a set, N, of readers associated with some portion of a supply chain. If a 100% read is achieved with $N = 1$, this is a condition of *absolute readability*. If $N > 1$, to achieve a 100% read rate, the condition is 100% *readability under extended conditions*. In simple terms, these conditions typically imply fast-moving consumer goods or pharmaceuticals, respectively, with two primary technology choices, namely between HF and UHF, and between near field and far field.

The fast-moving consumer goods (FMCG) and the initial pharmaceutical (PHARMA) industry pilots have taken different paths in their selection of UHF or HF, under possible variations of near-field or far-field RFID technologies to meet apparently different supply chain objectives. Business objectives for all companies considering RFID deployment include (1) product cost savings (impacting profit margins) versus (2) product security (managing corporate liability and image). A solution to these competing objectives involves the analysis of physics, new technology, legacy systems, and the human factors that serve specific company and industry objectives. This chapter will discuss these alternatives through the technologies chosen as opposed to the *boardroom* decisions that are well beyond the scope of this work.

4.2   Basics of the technologies

This initial discussion establishes several basic points regarding the choices between high frequency (HF) and ultra-high frequency (UHF), where the discussion will focus on 13.56 MHz for HF and 915 MHz for UHF. The choices to be made between these two alternatives are on the basis of the following factors (although the list may arguably not be complete): (1) physics, (2) technology, and (3) the human (commercial) factor.

In addition to HF and UHF, two additional *buzz words* are important: near field and far field. While an in-depth discussion of these two terms is beyond the scope of this chapter, a simple explanation will help shed some light on why both physics and technology are important. The fundamental choice of an implementation is really that of choosing one from column A and one from column B, where column A is (HF, UHF) and column B is (near field, far field).

In the area of passive RFID, it is necessary to transmit both energy and information to the RFID tag. Both energy and information originate with an interrogator (reader) that accomplishes such transmission using an antenna to interface to the air medium. The signal carrying both energy and information is sent to the antenna, at which point the transfer of energy into the air from the tag to the reader takes place. There are two primary mechanisms by which the energy is transferred to the tag, where in both cases the air interfaces are termed *antenna*, although one is more similar to the windings of an electrical transformer than to the classical image of an antenna.

Given an interrogator with an antenna, the near field is a reference to the particular technology in which the tag works relatively close to the antenna (something like an air core transformer) and far field is a technology for which the tag works further away from the antenna. The boundary between these two concepts is not well defined, but virtually all definitions are based on the maximum of any of the dimensions of the antenna. Conceptually, in the near field, the reader and the tag interact through an invisible coupling mechanism termed lines of flux (or simply flux). In the far field, the reader and the tag are not coupled, but rather, energy impinging on the tag antenna is reradiated back to the receiver of the reader, where this energy is termed *backscatter*.

4.2.1   Far field

Far field involves an antenna; typically the one used is either a patch or a single dipole (or some combination thereof) that radiates the energy from the interrogator antenna to the tag antenna (typically a dipole). In this case, the two antennas are not linked to each other by lines of flux as is the case with the near field. For efficient transfer of this energy taking advantage of the far field, the dimensions of the antenna need to be on the order of the wavelength of the signal frequency, i.e. 13.56×10^6 cycles per second (13.56 MHz) for HF and 915×10^6 cycles per second (915 MHz) for UHF. The product of the frequency, f, and the wavelength, λ, is equal to the speed of light, c, which is 3×10^8 meters per second, i.e. $c = f\lambda$. Therefore, the wavelength of UHF (915 MHz) is slightly more than a foot. The corresponding wavelength of HF (13.56 MHz) is over 67.5 feet. Thus, for a given amount of energy to be transferred from an HF antenna through the far-field mechanism, it would need to have dimensions on the order of 67.5 feet, which would be highly impractical for most supply chain applications. Thus, far-field transmission at 13.56 MHz is not practical in most supply chain applications. Alternatively, the near/far-field boundary according to one source is $D^2/(4\lambda)$ [1], where D is the maximum dimension of the antenna; say about 6 inches at 915 MHz. Thus, to operate in the near field at 915 MHz implies an operating distance of considerably less than a foot.

4.2.2   Near field

Near field, as mentioned previously, is a mode of operation involving a transformer-type action whereby the electrical signal applied to a winding of a transformer

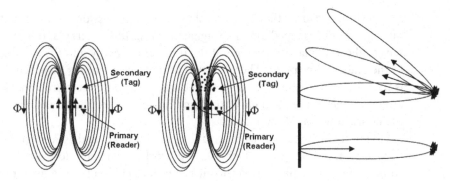

Fig. 4.1   Comparison of orientations for near field (NF) and far field (FF). (a) NF favorable
orientation; (b) NF variable orientation; and (c) FF tag (top), reader (bottom).

produces magnetic flux lines (field) that link the primary (interrogator) antenna to
the secondary (tag) antenna. The flux lines generated by the reader antenna are
shown in Figs. 4.1(a) and (b), from which it can be seen that these lines loop around
the reader antenna (primary). A tag is included in Fig. 4.1 where the tag antenna is
linked to the primary (reader antenna) through the flux lines.

   In summary, it is important to note that, in the far field, the energy that has left
the antenna does not return to the interrogator unless it is in some way reflected or
backscattered. In the near field, the energy from the interrogator antenna returns
to the antenna whether or not there is a tag in the field. If there is a tag, the
modulating impedance disturbs the field that is already coupled through a *return
path* to the interrogator antenna.

4.3   Fundamentals of orientation

Orientation variations between the near field and far field are best viewed by
considering how energy is delivered from the transmitter of the interrogator to the
tag. Far field implies energy transferred via the electric field, in which case the
orientation of the two antennas directly affects the backscattered energy by what
is termed *polarization*. Polarization implies that these antennas must have a specific relative orientation(s) for effective energy transfer. To compensate for mis-
alignment of the reader/tag antenna combination, some tags actually contain two
orthogonal (perpendicularly mounted) dipole antennas. In this case, the worst
possible alignment of the plane formed by the two-dimensional tag is 45° as
opposed to 90° otherwise. The trick is to efficiently connect the two antennas to
the chip input to maximize the benefit of both antennas. It is important to note
that the transfer (propagation) of electromagnetic energy involves two components, electric and magnetic fields, from which the name is derived. The magnetic
field will now be discussed as a part of the near field.

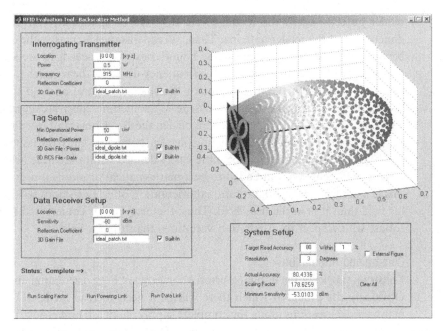

Fig. 4.2 A three-dimensional area of operation.

The near-field energy transfer involves a coupling of energy through flux lines in the magnetic field. Typically, a near-field antenna will have multiple turns in concentric circles forming a coil or inductor. The energy transferred is a function of the number of lines of flux encircled by the loops of the antenna (secondary). It is difficult to make generalized comparisons between near- and far-field orientation sensitivities without looking at specific antennas of both types. However, qualitatively speaking, it is felt that the near field with flux coupling, as shown in Figs. 4.1(a) and (b), is more robust with respect to acceptable orientations than the electric field energy in the far field. As indicated, this is a conjecture offered without proof at this time. The difficulty in proving such a conjecture is that the specific application must be taken into account in order to avoid contrary anecdotal examples.

Consider the near- and far-field illustrations of Fig. 4.1. The linking of the near-field flux lines through a tag as it is rotated, as shown in Fig. 4.1(b), illustrates the rotational nature of tag orientation while the lines of flux continue to flow through the antenna of the tag acting as the secondary of a transformer. In Fig. 4.1(c), the major beam that powers the tag in the first phase is shown passing from the reader to the tag at the bottom, where the tag is rotated with some variation as to the area available to harvest energy as a part of the powering phase of the read cycle. A variation of the form of the major beam for a specific antenna can also be seen in Fig. 4.2, which has been mathematically derived on the basis of the Friis equation [2]. The upper portion of Fig. 4.1(c) illustrates the energy patterns backscattered from the tag to the reader, where the result of a rotational orientation tends to direct the most favorable portion of the reflected energy in a path that is likely to miss the reader antenna when the reader-to-tag distance is significantly larger than the dimensions of the reader antenna. Thus, as read distance increases, the orientation becomes more critical. However, the boundaries of the near field are specified in terms of the physics involved with the frequency used, i.e. the wavelength, λ. In the case of the far field, the dimensions of the space of operation tend to be much greater, thus stressing the orientation issue. We sometimes illustrate the near field as a prisoner in a cell and the far field as a bird in the sky. Constraints tend to increase the probability of accurately predicting physical behavior. Thus, the issue of how to make the tag work better can be more easily defined within the near field.

This argument is not intended to be a mathematical or physical proof. Instead, this is a conjecture that the rotational orientation of a tag is more sensitive in the far field than it is in the near field. Thus, the near field is conjectured to be more robust with respect to orientation variations than the far field.

Constrained orientation

Constrained orientation implies a-priori knowledge of both orientation and position. Obviously the points at which there will be a need to read an RFID tag in the many varied locations along the complete supply chain are numerous. The totality of such orientations is beyond the scope of this work. However, a few anecdotal examples will serve to make the point. Consider a box (carton) of items on a conveyor belt of relatively small width. The carton forms three planes – ends, top/bottom, and sides. The carton has a small rotational variation in orientation in the linear direction of travel. Thus, there is considerable freedom in linear motion (translation) along the conveyor belt, but not in rotation, due to the side panels of the conveyor belt. The vertical height of the box is fixed and the maximum distance from the side of the conveyor belt is known and bounded by the width of the conveyor belt itself if only one reading antenna is used, since the box may be rotated by 180° from the nearest point to the reader antenna. Under such conditions, it is possible to use a barcode if the application of the label is automatic and consistent, along with two barcode readers in case the box is rotated by 180°. Thus, the RFID advantage here is that just one reader is needed instead of two if it is possible to read through the box (carton). As an aside, have you ever seen many (any) RFID applications take advantage of this fact?

It is possible to test the carton and the position as well as the orientation in a laboratory to insure that an operating RFID tag can be read in the constrained position and orientation discussed here. These tests merely require someone who knows how to operate the particular reader who has the carton dimensions and access to a model of the conveyor belt. Many companies and universities offer this type of service, although the user can do the same thing because they actually have the conveyor belt.

Unconstrained orientation

Unconstrained orientation implies no *a-priori* knowledge of the position and orientation of the RFID tag, in contrast to the case of the example carton in the previous section. However, it is assumed that the tag, in some orientation (usually the most favorable), is within the read range of the interrogator. The read range (distance) in general does not exist as an absolute number but rather as a three-dimensional space, as is shown in Fig. 4.2. This three-dimensional space will be termed the *envelope of functionality*. It must be noted that even within the surface outlined in Fig. 4.2 there are voids that simply cannot be read by a single reader, whether considered in the practical or the theoretical sense. This is a classical misconception in RFID. This statement is made on the basis of thinking of the tag as being in free space, such as an anechoic chamber. In certain practical applications, the environment may contain some type of reflecting element(s) that could overcome a no-read situation. This is really not a simple RFID reader/tag system. It now contains an additional element (reflector) that needs to be classified and included in the consideration.

To understand the situation, it is important to note that two conditions must occur in order to read a passive RFID tag as indicated above. First, the silicon chip on the tag must have sufficiently high levels of power and voltage to operate the electronics implementing the given protocol. This powering of the tag is the responsibility of the transmitting portion of the interrogator. Second, the receiver portion of the interrogator antenna must be able to "see" the modulated back-scatter response of the tag. The tag parameter affecting the ability of the receiver to "see" it is termed the *radar cross section* on the tag antenna. If the tag has sufficient voltage and power, the sensitivity of the receiver portion must be sufficient for it to be able to read the backscatter signal which is produced by the tag at the particular orientation.

The two electrical terms of power and voltage are important. The amount of power harvested is primarily a function of the antenna and the impedance matching to the chip. The required power and voltage are thus a function of the chip design and the fabrication technology. As technologies improve, both of these requirements are reduced. However, this is normally viewed as simply increasing the range, thus placing the same problem on the table, but at a longer range.

In the simplest of examples, assume a light (transmitting source of the interrogator) that illuminates an area in front of the fixture holding the light. A mirror placed in the illuminated area must be within a certain range of the light and within a certain range of angular orientations in order for it to be able to reflect the light back (*backscatter*) to be seen by the reader portion of the interrogator.

4.4 Antennas and materials

Independently of the operating field, maximizing the energy collected by the tag antenna is essential in order to optimize the operating range and robustness of passive transponders. In terms of classical electromagnetic theory, tag antennas are typically characterized by modeling or measuring antenna gain, radiation pattern, and input impedance. It is necessary to match the impedance of an antenna to a load in order to efficiently couple the RF power.

At present, RFID systems are used in identifying various products and containers with different electromagnetic properties. The dielectric properties and tangential losses for a particular product are varied and are distributed through the carton of multiple items due to the properties of different contents. The properties of an item-level product affect antenna impedance due to changes in the resonance (operating) frequency and matching between the antenna and the chip. Therefore, different antennas are designed in order to compensate for the detuning (negative) effects caused by different materials in the proximity of the antenna. In HF technology, there is a limited number of antenna variants (i.e. typically spiral inductors) with respect to the overall geometry. The typical antenna geometries include shapes such as circles, squares, and rectangles with different numbers of turns and different spacings between traces. This fact provides historical evidence of the more robust nature of HF–near-field RFID. This is in turn translated into a more robust solution across end uses during the tag selection process. By contrast, UHF technology provides a plethora of different antenna designs, which can be viewed as a "menu" and attributed to the need to understand and compensate for the varied use scenarios. Initially, tags can be grouped into three main categories: pallet-level, carton-level or item-level tags. Depending on the application, there are specific antenna types for products such as cartons, baggage, vials, and garments.

The antenna design process of UHF-type antennas normally requires a three-dimensional model of the product with the tag, which is accomplished with electromagnetic simulation software based on the finite element method (FEM). Any major errors translate into unread tags and come from inadequate specification of the distributed characteristics of the product and packaging materials. These issues are critical in the design process of UHF-type antennas, and are mainly due to the shorter wavelength used when operating in the far field as opposed to the wavelength used with HF-type antennas operating in the near field. The electromagnetic properties have much lower variation in these scenarios with reduced frequency.

The antenna material is another critical factor that affects HF and UHF technologies differently. Both technologies depend on the lowest possible resistance in the antenna material for both volume and sheet resistance, but UHF-type antennas are greatly affected by the *skin effect*. First of all, from the skin-effect standpoint, the antenna material is required to have an exceptionally smooth surface. Secondly, the UHF-type antennas are designed for the specific inductance value as part of the overall antenna structure that provides impedance matching for the chip. Therefore, the antenna characteristics typically depend on high-resolution designs with sharply defined edges. Also, the UHF technology is affected by the formation of oxide on the surface over time, which can degrade the performance due to an increase in the surface resistivity. Today, this problem can be resolved by using non-traditional materials such as conductive polymers.

4.5   An analogy to network layering

While there are obviously many different applications of RFID, it is quite clear from the basic idea of radio communication with a *thing*, and a back-end communication to a network, that there is a level at which all RFID implementations can be viewed as a simple layered structure similar to the classical ISO layering that was originally developed for wired networks. The advantage of such a layering structure is that the individual layers, software and/or hardware, can be interchanged depending on the particular application, provided that the interfaces connecting adjacent layers are sufficiently well designed. Eventually, although this is not being proposed at this time, the physical layer front end could be barcodes, infrared (IR), bluetooth, etc. However, for the conceptual development here, the proposed physical layer will be either 915 MHz or 13.56 MHz.

The two primary interfaces of an interrogator are the network (the back end) and the air interface. The back end is important technology that connects the interrogator to the network (world). Such connections are commonplace in information technology because of the numerous different types of devices that are network accessible. The physical connection at the back end may be wired or wireless.

Networked data collection and communication devices can have a variety of physical (air) interfaces, including RF and IR. Thus, there is no technological or physical reason why both UHF and HF cannot be interchangeable physical layers for either a classical UHF or an HF RFID back end. There are two limitations that could potentially affect the established theoretical ability to interchange these facets. (1) The specified data rate at the physical layer; and (2) the existence of the required information on/in the device to be interrogated.

The data rate is a specification that may limit the above solution because of physical constraints due to the HF frequency range. Thus, it does not limit the physical reality, but only some range of data rates within that reality. The primary show-stopper is the availability of the required information in one embodiment or the other (UHF or HF). This is not a physical limitation but rather one of standards, specification or personal accommodation.

OSI layering
Basic layering for the OSI [3] [4] seven-layer model is shown in Fig. 4.3(a), where the connection to the internet is through the application layer with the physical layer being the interface to the medium. The OSI layering allows interoperability between different components because a component developed for one layer need only adhere to the OSI interfaces to the layer above and the layer below. For example, the medium can be a hardwired interface or an air medium.

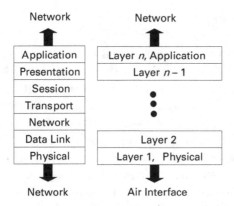

Fig. 4.3   (a) The OSI seven-layer model; (b) an RFID *n*-layer model.

Obviously, layer 1, the physical layer, can be implemented with either UHF or HF. In fact, LF could also be used, although the data rate would be lowered considerably. However, this is not a physical or a technological limitation. Given that the object being interrogated contains the required information, the application layer *n* is simply a manner of formatting and programming at a high level of programming. Layer 2 can simply be the classical media-access control with, say, layer 3 handling the protocol for a particular standard or technology. The establishment of the functionality and the implementation of the individual layers is not *rocket science*, but it does involve collaboration and accommodation, which are perhaps more difficult than *rocket science*.

Physical layer wireless interface specification

Network layering for passive (active) RFID can be implemented in the same form with the corresponding blanks to be filled in as for Fig. 4.3(b). The fundamental issues to be dealt with concern the air interface or physical layer. Using the HF/UHF alternative, two alternative **physical layers** are called for, with a variation in the antenna implemented in each case. In the simplest sense, the next layer is the **media-access control** that is based on the particular protocol, where the primary consideration may be collision avoidance. Otherwise, in the case of a single tag, there is little complication. This is in no way intended to trivialize the collision problem. As a first estimate, the **protocol layer** would be layer 3. The intent here is not to define the recommended layers but rather to provide initial insight into the process of differentiating functionality so as to provide for interchangeable layers that can be designed independently. Thus, the most important aspect is the definition of the functionality of the individual layers, along with the interfaces between layers.

This concept is independent of active or passive implementations. The only condition is that the tag must contain the necessary identification required by the back end of the reader at the network interface.

4.6   Examples of converging technologies

At any given point in time, hardware and software products are under development, as are standards and commercial applications. Companies have hardware
products that are (1) on the market, (2) about to be released, (3) under development, and (4) at various levels of research. The product lines are based on the
corporate view of the situation and the market at the point when decisions must
be made. A change in standards and new technology developments cause disruption in the commercial market as well as in individual companies. Thus, the
tendency in change is to move slowly in order to avoid major disruptions and push
back at various levels. However, in rapidly expanding markets, change is inevit-
able and is recognized by all of the players.

Companies have representation on various standards bodies and they try, within
reason, to influence the standards so as to avoid total disaster for their individual
companies. Typical standards groups involved in RFID have been the ISO and the
ANSI. The major excitement in RFID has obviously been brought about by the
Wal-Mart initiative which has resulted in the current specification known as
EPCglobal Class 1 EPC GenII (EPC GenII), for which a number of companies are
supplying silicon and numerous other companies are providing inlays and tags
under this specification. Recent efforts have incorporated the EPCglobal specification into the ISO 18000 set of RFID standards as ISO 18000 – Part 6c.

4.6.1   EPCglobal Class 1 EPC GenII/ISO 18000 Part 6c

This protocol has exhibited success with a variety of first adopters, including pilot
projects. The fundamental protocol involves (1) a query command by the reader,
(2) a 16-bit pseudo-random-number response from a tag, (3) an echo of the 16-bit
number by the reader, and (4) a response by the tag sending the EPCglobal 96-bit
number. This exchange can be seen in Fig. 4.4.

The EPC exchange is accomplished repeatedly in the far field at distances of
tens of feet with success.

4.6.2   ISO 18000 Part 3

This near-field standard is being designed for item-level tagging to comply with
the 100% read requirements for verifying pharmaceutical totes.

4.7   Technical summary

The ultimate choice of a frequency, HF or UHF, is dependent on many factors
that often miss being considered in the decision as to which choice is the most
appropriate for a given application. As a part of many discussions regarding HF

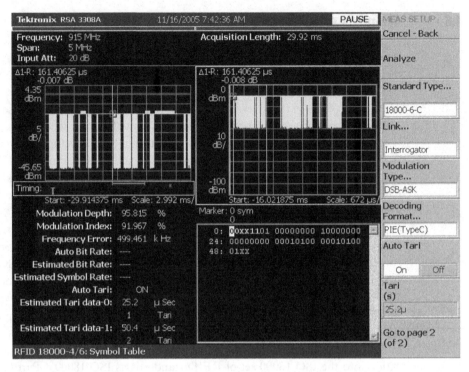

Fig. 4.4   EPCglobal Class 1 GenII query command and tag/reader exchanges.

and UHF, the terms near field and far field are used without a full understanding of what is involved. The choice of HF or UHF is coupled to a choice of near field and far field giving a set of four possibilities for a given application. These choices depend on factors including the distance(s) of operation and all possible orientations of the item to be tagged.

While the optimum choice can be determined by a careful analysis of how the tagged item will be oriented when read, the distance involved and the robustness across an entire production line and across the supply chain are parameters that are not easily or inexpensively obtained. The important point is that a choice must likely be made without the best evidence for the application. Practically, it becomes important to have a tag to apply to the item at a given point in time. If immediate application is required, the best decisions can often be made by creating realistic conditions in a pilot project and objectively weighing the levels of performance of various alternatives deploying commercially available technology. These alternatives and the parameters of the application are likely not strongly correlated. It is this dilemma that makes pilot projects an excellent resource for resolving practical issues. While the ultimate choice might not be made with all possible information, systematically gathering the data and carrying out subsequent analysis in a more exhaustive survey may be more expensive and less effective than a simple pilot project that uses each of the four choices with accurate and detailed reporting of the results of the read rates correlated with the operational scenarios.

4.8   Pharma – a surrogate for the future

The initial impetus for RFID deployments came from the Wal-Mart initiative for pallet and carton tagging with UHF for fast-moving consumer goods (FMCG). This initiative has been extremely important to the advancement of RFID technology in the interest both of vendors and of suppliers. However, it has not reached the point of providing a return on investment (ROI) for many suppliers. The threat of lost business keeps the major suppliers committed in spite of the current costs of tagging products.

RFID has shown ROI in numerous anecdotal examples where the cost of goods is sufficiently high to mitigate necessary compromises and handling restrictions of "closed loop" stand-alone alternatives not available in the FMCG supply chain applications.

The more immediate mandate is that of Federal Drug Administration (FDA)-inspired regulations in states such as Florida and California that mandate ePedigree tracking for consumer drug sales that reports chain-of-custody information from supplier(s) across all parties in the supply chain to the end user. While potential profit margins as a result of FMCG supply chain visibility are certainly economic incentives, the liability imposed by organizations such as the FDA is incentive enough to choose a technology based on compliance. This choice has seen early adopters choose HF near field as opposed to trying to support the technology initiatives in UHF.

The pharma industry is an excellent industry for RFID pilots and provides numerous opportunities for RFID experimentation and optimization. The industry provides diversity in the supply chain due to the nature of the commercial outlets, which are relatively small, compared with Wal-Mart, and quite numerous. Owing to the relatively small size of the commercial outlets, in order to provide the maximum number of different products on the shelf, the quantity of stock is small. The point-of-sale (POS) information is used to restock the shelves from a relatively small distribution center (DC), with the quantities of product typically smaller than a full carton. Shipments from the DC to stores are assembled in mixed totes. The mixed-tote method of delivery requires the ability to read items with 100% accuracy just to get from the DC to the store. However, at some point in the supply chain, cartons must be shipped, providing supply chain diversity not seen in other areas of FMCG.

Security/privacy is another factor required in the pharma sector. This is primarily due to the privacy issues involved with certain medications, in particular those associated with HIV, for which the chain of custody must be maintained while providing privacy on final delivery. However, privacy is a two-edged sword in that, while it is possible to "kill" a tag, for example by using the EPC GenII protocol, killing the tag forfeits any future benefit of the RFID tag for tracking purposes when discrepancies are encountered. While killing the tag may be important to privacy, it also introduces difficulties if there is a problem with the medication after it has been delivered. In order to access the chain of custody, it is necessary to read the tag to determine the fault in the chain.

This chapter has covered the near-field and far-field characteristics of specific unlicensed frequencies and the orientation restraints of RFID systems, together with proposing a layered approach to architecting RFID protocols that is independent of frequencies, to help work through the systems design issues for multi-frequency RFID installations.

4.9   References

[1] **US Federal Communications Commission, Office of Engineering and Technology**, "Evaluating Compliance with FCC Guidelines for Human Exposure to Radiofrequency Electromagnetic Fields," *OET Bulletin 65*, pp. 3, 27–30 (August 1997).

[2] **Finkenzeller, K.**, *RFID Handbook: Fundamentals and Applications in Contactless Smart Cards and Identification* (Wiley, New York, 2003).

[3] **Stallings, W.**, *Data & Computer Communications*, 6th edn. (Prentice-Hall, Upper Saddle River, NJ, 2000).

[4] **Forouzan, B. A.** and **Fegan, S. C.**, *TCP/IP Protocol Suite* (McGraw-Hill, New York, 2003).

5 Integrating sensors and actuators into RFID tags

J. T. Cain and Kang Lee

5.1 Introduction

As applications of RFID tags and systems expand, there is an increasing need to integrate sensors into RFID tags. The RFID tags being used at present in the supply chain indicate what a product is, but do not reveal any information about conditions that the product has encountered throughout its passage along the supply chain. Only a few RFID tags with sensors are commercially available and they are custom-designed tag–sensor combinations. Adding sensors that can measure environmental conditions such as temperature, vibration, chemicals, gases, and health, and the capability to interrogate the sensor outputs, can provide much needed information about the current and historical conditions of the product. The ability to incorporate sensors and possibly actuators into RFID tags would also open a whole new world of imaginable applications in homeland defense, military operations, manufacturing, animal health, medical operations, and other applications. The IEEE (Institute of Electrical and Electronics Engineers) 1451 suite of standards [1–9] was developed to provide for "smart" transducers (sensors and actuators) and flexible network interfaces that facilitate "plug-and-play" capabilities for the transducers. The objective of this chapter is to present the current situation in RFID systems and networked transducers and to describe the strategy that is being adopted and research that will be necessary in order to incorporate "smart" sensors and actuators into existing RFID tags and systems using the IEEE 1451 suite of standards approach.

5.2 RFID systems

A basic RFID system consists of an RFID tag or a set of multiple tags and an interrogator or reader as shown in Fig. 5.1. In most present-day applications, a tag is attached to an item that is to be identified or tracked. The interrogator or reader communicates with tags that are within its range. The reader controls the communications

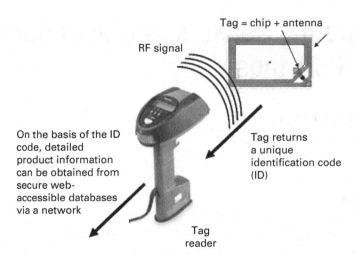

RF signal

Tag = chip + antenna

On the basis of the ID code, detailed product information can be obtained from secure web-accessible databases via a network

Tag returns a unique identification code (ID)

Tag reader

Fig. 5.1   A basic RFID system.

with the tags. Basically there are three types of tags: active tags, semi-passive tags, and passive tags. Active tags and semi-passive tags contain an energy source, normally a battery. Passive tags contain no power source and must convert energy from the RF signal provided by the interrogator in order to operate the on-board electronic chip. The tag is able to send back data stored on the chip. RFID systems and tags are also distinguished by the frequency spectrum where they operate.

The active tags that we consider here are governed by the ISO/IEC (International Organization for Standardization/International Electrotechnical Commission) 18000-7:2004 (referred to as Part 7) standard. This standard defines the air interface for readers and tags operating in the 433-MHz band – an ultra-high-frequency (UHF) band. Currently this standard, together with other standards in the ISO/IEC 18000 series of standards, forms the technical basis for container security and tracking applications.

Systems that comply with this standard will have the ability to identify any tag within a range of 1 meter to 100 meters, write and read data to and from the tag, select tags by group address or individual tag addresses, handle multiple tags within the range, and carry out error detection. The standard specifies the physical layer, including the carrier frequency, modulation type, and modulation rate. It also specifies that the interrogator–tag relationship is master–slave.

The data link layer is also specified. The packet format is specified as a preamble followed by data byte(s) and then CRC (cyclic redundancy check) bytes. The standard also specifies the interrogator–tag and tag–interrogator message formats and the command codes. Also specified by the standard is a collision detection and arbitration scheme that allows successful querying of multiple active tags within the range of the interrogator.

The passive tags to be considered here are governed by the ISO/IEC 18000-6:2004/Amd 1:2006 standard (colloquially known as the ISO/IEC 18000-6c standard, incorporating the EPCglobal Class-1 Generation-2 UHF standard), which specifies the air interface both for battery-assisted and for completely passive tags. In this case, the carrier frequency for the RFID communication protocol is specified to be in the 860 MHz–960 MHz range (designated as the ISM (industrial, science, and medical) band in the USA) covering the range from 1 meter to 30 meters. This is because different countries around the globe have allocated different frequencies in this range of the UHF band for reader–tag communications. In a passive tag system, the interrogator initiates communications by sending out RF signals. The tag then communicates to the interrogator via backscatter, i.e. the tag responds by modulating the reflection coefficient of its antenna, producing a series of binary RF signals that correspond to ones and zeros that comprise the response. Tags in conformance with this standard must have the ability to store an electronic product code and a tag identifier and implement a "kill" command that permanently disables the tag when issued by the interrogator. Other options include password-protected access control and writable memory.

As with the active tag, the standard specifies the physical layer, the data link layer, and the collision arbitration scheme used to identify a specific tag in a multiple-tag environment. Both layers are more complex than that of the ISO/IEC 18000-7:2004 standard, but all parameters of the physical layer are defined and the data link layer includes the format and definition of interrogator–tag and tag–interrogator protocols and commands. Again, there is additional complexity over Part 7 that requires state machines to be defined as a part of the standard.

Proprietary active tags that include sensors have been produced. Naturally, it is more challenging to incorporate sensors or actuators into passive tags because of the limited energy available for sending data. There is a need for a standard or series of standards to address the integration of tags with sensors and/or actuators. Work has begun at the IEEE sensor standards committee and ISO/IEC joint committee on RFID systems to define standards to encourage interoperability or ease of adding/changing sensors and/or actuators for RFID tags. In the following section we will review the status of standards for networked "smart" sensors and actuators. These sensor standards are quite relevant for integration with RFID systems.

5.3   "Smart" transducers

The IEEE Instrumentation and Measurement Society's Technical Committee on Sensor Technology TC-9 has developed a suite of standards for the design and implementation of "smart transducers" (the IEEE 1451 suite of standards). The goals of the IEEE 1451 suite of standards are as follows.

- To provide network-independent and vendor-independent transducer (sensor or actuator) interfaces.
- To provide a standardized format for transducer electronic data sheets (TEDS) that contain manufacturer-related data for transducers.

- To support a general model for transducer data, control, timing, configuration, and calibration.
- To allow transducers to be installed, upgraded, replaced, or moved with minimal effort.
- To eliminate error-prone manual entry of data and system configuration steps – thus achieving "plug-and-play" capability.
- To allow wired or wireless sensor data to be moved seamlessly to/from the network or host system.

A top-level view of the IEEE 1451 concept is shown in Fig. 5.2. There are two principal components in an IEEE 1451-based system: the network capable application processor (NCAP) and the transducer interface module (TIM).

As the name implies, the NCAP is assumed to be the unit where most of the software implementing the application-specific functions is located. The NCAP is covered by the IEEE 1451.0 and IEEE 1451.1 standards. These standards enable an object-oriented consistent application programming interface (API) environment to be independent of the type of network to which the NCAP is connected. The IEEE 1451.0 standard defines the common commands, TEDS, and functionality for the IEEE 1451 suite of standards. It allows the access of IEEE 1451 sensors and actuators via wired or wireless interfaces defined by each of the IEEE 1451.X physical layers described later. On the other hand, the IEEE 1451.1 standard defines the object model of smart transducers, the behavior of smart transducers, the architecture of the interface to the network, and the client/server and publish/subscribe methods for communication with other NCAPs.

The TIM, as the name implies, provides the interface to the transducers. A TIM can have any mix of a maximum of 255 transducers. Individual transducers can be

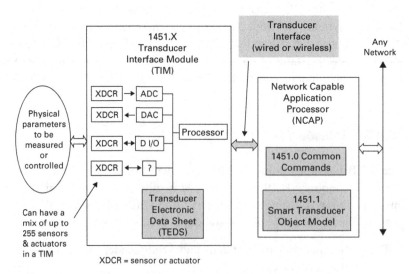

Fig. 5.2   Top-level view of the IEEE 1451 concept (from prior publication in IEEE standard 1451. NIST, SH95225, 2004, with permission).

sensors that produce an analog output requiring an analog-to-digital converter (ADC) for signal conversion or sensors that produce a direct digital output. Other direct digital inputs and outputs are also possible. Actuator outputs can also be either analog, requiring a digital-to-analog converter (DAC), or direct digital outputs.

A key aspect of the suite of standards is the inclusion in the TIM of the TEDS. This is the feature that enables most of the goals of the suite to be achieved. The TEDS enable

- self-identification and self-description of sensors and actuators;
- facilitation of sensor system configuration;
- diminution of human error in manual system configuration;
- simplification of field installation, upgrading, and maintenance of sensors by simple "plug and play" of devices to instruments and networks;
- self-documentation;
- enhancement of support for conditioned-based maintenance (CBM) systems.

A TEDS contains manufacturer-related information about the sensors and actuators. The IEEE 1451 suite of standards defines different types of TEDS to meet the needs of various industry sectors, some of which are the following.

- **Meta-TEDS**
  - Provides all the information needed to gain access to any transducer channel and information common to all transducer channels in a TIM such as a universally unique identifier (UUID), how to access meta-TEDS, and the number of transducer channels.
- **Transducer channel TEDS**
  - Provides the characteristics of a single transducer channel.
  - Specifies lower and upper range limits, timing, uncertainty, physical units, update time, and self-test.
- **PHY TEDS**
  - Defines parameters unique to the physical communications media used to connect the TIM to the NCAP.
- **Calibration TEDS**
  - Provides the calibration constants to convert the output of a sensor into engineering units and vice versa for an actuator.
  - Specifies last calibration date–time, calibration interval, and multinomial coefficient.
- **Geographic location TEDS**
  - Provides location of the sensor on the basis of OGC (Open Geospatial Consortium) geo-location specifications.
- **Frequency response TEDS**
  - Provides the frequency response data for a single transducer channel as a table.
- **Transfer function TEDS**
  - Provides the frequency response data for a single transducer channel with an algorithm.

- Allows user to combine this with the desired response to compensate the data.
- **Manufacturer-defined TEDS**
  - Allows the manufacturer to define additional features.
- **Text-based TEDS**
  - Allows manufacturers to provide textual information with the device, specified in the language written.
  - Specifies that a structure containing text is written in XML (eXtensible Mark-up Language).
- **End-user application-specific TEDS**
  - Allows user to write application-dependent data that the user keeps with the TIM or transducer channel.

Many of the above TEDS are transducer- and application-dependent. In order to minimize the requirements for the standards, many TEDS are set to be optional. The first three are essential to the goals of the suite of standards and thus they are mandatory.

Another key aspect of the suite of standards is the interface between the TIM and the NCAP. As indicated in Fig. 5.1, this interface can be wired or wireless. There is a separate standard (an IEEE 1451.X) in the suite for each unique interface included to date. Figure 5.3 shows the current set of interface standards in existence or under development.

The initial TIM standard was IEEE 1451.2. It specified a 10-wire interface between an NCAP and a TIM. It is currently in revision to include a universal

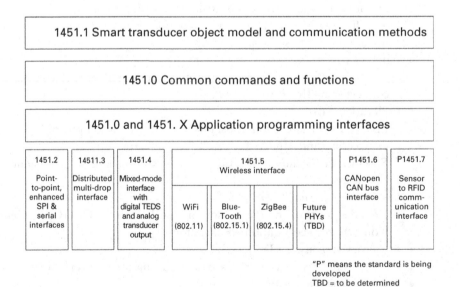

Fig. 5.3   The IEEE 1451 suite of standards (from prior publication in IEEE standard 1451. NIST, SH95225, 2004, with permission).

asynchronous receiver/transmitter (UART) interface in addition to the existing enhanced serial peripheral interface (SPI) specification. This standard interface enables the "plug and play" of transducers in a network.

IEEE 1451.3 specifies a distributed multi-drop, bus-oriented interface that can accommodate multiple TIMs with a single NCAP. This standard needs to be revised to work with the newly approved IEEE 1451.0 standard.

IEEE 1451.4 is a standard originally defined to accommodate existing analog transducers with the addition of the TEDS to form transducer systems operated in mixed mode (digital and analog). In operation, the TEDS in the transducer are digitally transmitted to the host system or network via a wired interface and protocol. Then the transducer is switched to output its analog signals to the host system via the same wired interface. The conversion and processing of the analog transducer signals by the host system are not defined in the standard.

IEEE 1451.5 is a wireless transducer interface standard. Its main objective is to provide data-level interoperability for sensors and actuators by combining the benefit of the TEDS and the adoption of existing popular wireless communication protocols in the standard. The standard is open to include other applicable wireless communication protocols; some possible candidates are the Ultra Wide Band (UWB) and WiMax (Worldwide Interoperability of Microwave Access) wireless protocols. To date it has included the following protocols:

- IEEE 802.11x – the WiFi (Wireless Fidelity) standard,
- IEEE 802.15.1 – the Bluetooth standard, and
- IEEE 802.15.4 – the ZigBee standard.

IEEE 1451.6 is intended to define a sensor interface standard for CANopen-based networks incorporating the IEEE 1451 TEDS concept for intrinsically safe and other applications. This proposed standard allows the development of CANopen-based gateways and cascaded transducer networks that can use IEEE 1451 transducers.

IEEE 1451.7 is a proposed sensor-integrated RFID tag standard for battery-assisted tags. It is intended to define communication methods and data formats for sensors communicating with a RFID tag that follow the ISO-24753 standard. The IEEE 1451.7 standard specifies a physical or wired interface for connecting sensor(s) to a battery-assisted RFID tag and methods to report sensor data within the existing RFID tag/system infrastructure. The physical interfaces such as SPI and inter-integrated circuit (I2C) bus are being considered as the sensors for an RFID tag interface. Sensor data are transferred to the RFID tag and then to an interrogator/scanner upon scanning via the ISO-18000 air interfaces. The communication methods are intended to work with battery-assisted tags and with passive tags as well. This standard is not intended to cover sensor-integrated active tags that possess mesh networking capability, for which the IEEE 1451.5 standard is more suitable.

This suite of IEEE 1451 standards has proven to be very valuable in facilitating networked "smart" transducers. There continues to be interest in expanding the suite to include additional interfaces taking advantage of emerging technologies.

5.4   RFID tags with sensors

As noted, if sensors and possibly actuators can be integrated into RFID tags and the interrogators in a manner that meets the goals of the IEEE 1451 suite, the needs of current RFID systems for cold chains could be met and new applications of RFID would be enabled. Exploratory talks between appropriate IEEE and ISO members to examine the compatibility between the IEEE 1451 standards and ISO RFID standards are going on.

The main question is "Can IEEE 1451 be a basis for RFID systems with tags that include sensors and possibly actuators?" We believe that the answer is yes, but a number of issues must be addressed.

As described earlier, the basic components in an RFID system are the interrogator/reader and the tag. For the IEEE 1451-based sensor system, the basic components are the NCAP and the TIM. One immediate possibility is to merge the RFID interrogator/reader with the IEEE 1451 NCAP and the RFID tag with the IEEE 1451 TIM as shown in Fig. 5.4.

Naturally, passive-tag-based systems provide more of a challenge because of the energy/power constraint. Passive tags that incorporate temperature sensors have been developed at the University of Pittsburgh [10], but they were not designed in accord with either the IEEE 1451 or the EPCglobal GenII standards. Commercially available tags incorporating sensors are mainly active or semi-active tags.

One specific issue is whether the TEDS structure outlined above has all the information that is applicable to sensor-integrated RFID systems. In addition,

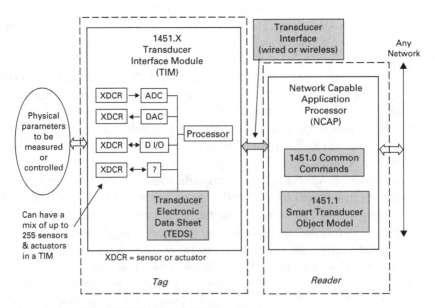

Fig. 5.4   Association of the IEEE 1451 and RFID systems.

another issue is whether the IEEE 1451.5 concept works well with the ISO RFID standards. When attempting to design and implement both the interrogator/reader NCAP and the tag TIM as units satisfying both sets of standards, tests need to be conducted to determine whether IEEE 1451.0/1451.5 are compatible with ISO/IEC18000-7 or 18000-6(c).

After careful examination by the ISO and IEEE working groups, it was determined that the present TEDS defined in the IEEE 1451 standards could not meet the needs of the ISO/IEC RFID standards for sensor integration. In addition, the ISO committee placed two constraints for sensor integration with the ISO/IEC 18000 series of standards: (1) the sensor has to work with existing RFID communication protocols and air interfaces defined in the ISO/IEC 18000 standards, and (2) the sensor has to work with the limited amount of data memory in the tag and data communication rate between the tag and the reader. Working within these constraints, a scheme for integrating one or more sensors and an RFID tag is proposed as shown in Fig. 5.5.

If it turns out that IEEE 1451.5 is too much of a constraint, then a new member of the IEEE 1451 family, IEEE 1451.x, could be developed to eliminate any conflict that precludes utilizing IEEE 1451.5. The test structures could be modified to test and evaluate interrogator/reader NCAP and tag TIM structures against the newly developed standards in the IEEE 1451 suite of standards.

This type of experimental design, implementation, and testing is well suited to the university research environment. For example, at the University of Pittsburgh, a development suite that enables rapid design and implementation of readers and tags compliant with ISO 18000 Part 7 has been developed and is

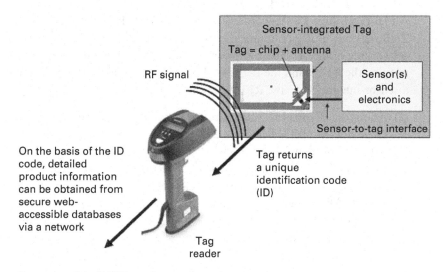

Fig. 5.5   Sensors and an RFID tag integration scheme.

close to completion with a similar system that would do the same for readers and tags compliant with ISO/IEC 18000-6c. Research at the NIST will continue to support the development of suitable interface standards for integrating IEEE 1451-based sensors with RFID tags and systems to meet industry needs.

An IEEE Sensor and RFID Integration Working Group based on the approach shown in Figs. 5.4 and 5.5 was formed in April 2007 to initiate the development of an IEEE 1451.7 standard for interfacing sensors to a battery-assisted RFID tag and passing sensor and RFID information to the interrogator/scanner via the existing ISO 18000 air interfaces. The key items to be defined in the standard are the logical and physical interfaces between the sensors and the tag, TEDS for the sensors, sensor data format in the tag memory, and the network access of sensor and tag data via a network and the web. All these are based on the IEEE 1451 sensor standards. The working group aims to complete the IEEE 1451.7 standard in 2008.

On another front, there is significant academic and industry interest in combining the concepts of wireless mesh sensor networks and active RFID systems. The reason is that, in a collaborative business environment, the tags and sensors on products would be able to communicate and share data with each other for improving efficiency and effectiveness of product-condition monitoring and management. Thus applying wireless mesh sensor networking solutions to RFID systems would create a new generation of sensor-based active RFID systems. Taking advantage of mesh sensor networking systems, each RFID-based sensor node can collect sensor data and transmit it to any other node in the network. Each node transmits not only its unique ID tag number but also details of its product condition and environment. The distributed sensor architecture of IEEE 1451 allows each NCAP to communicate with each other; therefore the IEEE 1451.5 wireless sensor standard is quite suitable for implementing the active RFID system concept described.

Figure 5.6 shows a view of IEEE 1451.5. Since IEEE 1451.5 adopts the popular ZigBee as one of its communication protocols and ZigBee can operate in a mesh-networking mode, it is mesh-network-enabled. Thus there are two possibilities for merging the RFID and the IEEE 1451.5 wireless sensor standards to come up with a sensor-integrated active RFID standard.

(1) Adopt the IEEE 1451.5 standard for active RFID systems by adding a tag ID to each wireless TIM node (see Fig. 5.7).
(2) Add an additional block under "Future PHYs" for RFID systems, such as EPC GenII physical layer (PHY) connections. The physical layer and data link layer in the two standards would define the interface. The flexible IEEE 1451.0 API structure in the NCAP would facilitate incorporating the interrogator/reader commands into the NCAP structure. Nevertheless, these two basic ideas need to be further investigated for feasibility and meeting the needs of industry.

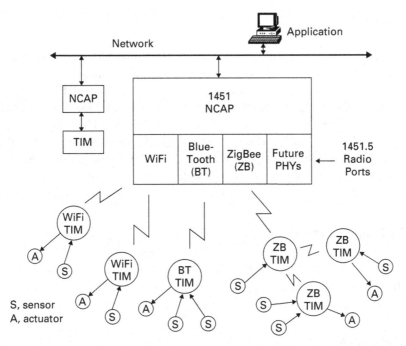

Fig. 5.6   A view of IEEE 1451.5 transport-independent sensor and actuator interfaces (from prior publication in IEEE standard 1451. NIST, SH95225, 2004, with permission).

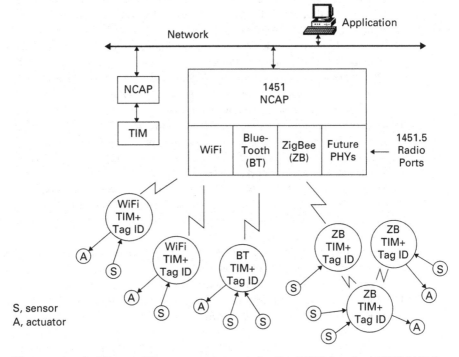

Fig. 5.7   The concept of adding tag IDs to sensor-integrated active RFID based on IEEE 1451.5.

5.5   Conclusion

It is clear that, in the future, the integration of sensors and likely actuators with RFID tags will be required to meet the needs of supply chains worldwide. We believe that the merging of the standards from both the ISO/IEC RFID domain and the IEEE smart transducer domain is a possible step in addressing this requirement for integration. Steps have been taken in establishing the IEEE 1451.7 standard for battery-assisted, sensor-integrated RFID tags. More investigation will take place to pursue the development of sensor-based active RFID systems and standards that will take advantage of the capability of wireless mesh sensor networks.

5.6   Acknowledgment

The author thanks the IEEE 1451 working groups for the use of the materials for this chapter.

5.7   References

[1] **Lee, K.,** "IEEE 1451: A Standard in Support of Smart Transducer Networking," *Proceedings of the IEEE Instrumentation and Measurement Technology Conference (IMTC)*, Baltimore, MD, Vol. 2 (National Institute of Standards and Technology, Gaithersburg, MD, 2000), pp. 525–528.
[2] **Wiczer, J.,** and **Lee, K.,** "A Unifying Standard for Interfacing Transducers to Networks – IEEE 1451.0," *Proceedings of ISA Conference*, Chicago, IL (2005).
[3] **IEEE,** *Standard 1451.1–1999, IEEE Standard for a Smart Transducer Interface for Sensors and Actuators – Network Capable Application Processor Information Model* (Institute of Electrical and Electronic Engineers, New York, 2000).
[4] **IEEE,** *Standard 1451.2–1997, IEEE Standard for a Smart Transducer Interface for Sensors and Actuators – Transducer to Microprocessor Communication Protocols and Transducer Electronic Data Sheet (TEDS) Formats* (Institute of Electrical and Electronic Engineers, New York, 1998).
[5] **IEEE,** *Standard 1451.3–2003, IEEE Standard for a Smart Transducer Interface for Sensors and Actuators – Digital Communication Protocols and Transducer Electronic Data Sheet (TEDS) Formats for Distributed Multidrop Systems* (Institute of Electrical and Electronic Engineers, New York, 2004).
[6] **IEEE,** *Standard 1451.4–2004, IEEE Standard for a Smart Transducer Interface for Sensors and Actuators – Mixed-mode Communication Protocols and Transducer Electronic Data Sheet (TEDS) Formats* (Institute of Electrical and Electronic Engineers, New York, 2004).
[7] **IEEE,** *Standard 1451* (NIST, Washington, DC, 2004) (http://ieee1451.nist.gov).
[8] **IEEE,** *Standard 1451.0–200??, IEEE Standard for a Smart Transducer Interface for Sensors and Actuators – Mixed-mode Communication Protocols and Transducer Electronic Data Sheet (TEDS) Formats* (Institute of Electrical and Electronic Engineers, New York, 2004) (http://grouper.ieee.org/groups/1451/0/).

[9] **IEEE**, *Standard 1451.5–2004, IEEE Standard for a Smart Transducer Interface for Sensors and Actuators – Mixed-mode Communication Protocols and Transducer Electronic Data Sheet (TEDS) Formats* (Institute of Electrical and Electronic Engineers, New York, 2004) (http://grouper.ieee.org/groups/1451/5/).

[10] **Mickle, M. H.**, **Cain, J. T.**, **Minhong Mi**, and **Minor, T.**, "A Circuit Model for Passive RF Autonomous Devices with Protocol Considerations," *International Journal of Computers and Applications*, 28(3):243–250 (2006).

6 Performance evaluation of WiFi RFID localization technologies

Mohammad Heidari and Kaveh Pahlavan

In the chapter that follows a test methodology is proposed for evaluating active RFID systems performance with wireless localization technology. While the problem of locating objects has been largely addressed for outdoor environments with such technologies as GPS, for indoor radio propagation environments the location problem is recognized to be very challenging, due to the presence of severe multipath and shadow fading. Several companies are now developing products to use RFID technology together with traditional localization techniques in order to provide a solution to the indoor localization problem. However, the performance of such systems has been found to vary widely from one indoor environment to another. A framework and design for a real-time testbed for evaluating indoor RFID positioning systems is described.

Historically radio direction-finding is the oldest form of radio navigation. Before 1960 navigators used movable loop antennas to locate commercial AM stations near cities. In some cases they used marine radiolocation beacons, which share a range of frequencies just above AM radio with amateur radio operators. LORAN (LOng RAnge Navigation) systems also used time-of-flight radio signals, from radio stations on the ground, whereas VOR (VHF omnidirectional range) systems in aircraft use an antenna array that transmits two signals simultaneously. The UWB two-way time transfer technique allows even more accurate calculation of distances for location tracking. Choices for in-building tracking systems are covered in the following chapter, including angle of arrival (AOA), received signal strength (RSS), time of arrival (TOA), and time difference of arrival (TDOA).

Localization and active tracking are some of the most promising commercial areas for active RFID suppliers, as the recent acquisitions of SAVI Technology and Wherenet demonstrate. Location-tracking applications are used to find people as well as equipment in hospitals, and across defense logistics, automotive, and other high-value supply chains.

RFID Technology and Applications, eds. Stephen B. Miles, Sanjay E. Sharma, and John R. Williams. Published by Cambridge University Press. © Cambridge University Press 2008.

6.1   Introduction

Localization using radio signals has attracted considerable attention in the fields of telecommunication and navigation. The very first program to address this problem on a worldwide scale resulted in the launching of a series of satellites for the Global Positioning System (GPS) [1]. Although GPS is widely used today for personal and commercial outdoor applications in open areas, it is recognized that it does not perform satisfactorily in indoor areas. Predicting the location of an individual or an object in an indoor environment can be a difficult task, often producing ambiguous results, due to the harsh wireless propagation environment in most such areas. The indoor radio propagation channel is characterized as site-specific, exhibiting severe multipath effects and low probability of line-of-sight (LOS) signal propagation between the transmitter and receiver [2], making accurate indoor positioning very challenging.

Radio frequency identification (RFID) is an automatic identification technology that relies on storing and remotely retrieving data using devices called RFID tags. The RFID tags are essentially radio transponders [3] [4]. There has recently been an enormous amount of innovative development of RFID devices, and the emerging RFID devices are being integrated with traditional localization techniques, especially in real-time location systems (RTLS) [5]. As expected, the new RFID technologies have given rise to a multitude of applications in commercial, healthcare, public safety, and military domains. In the commercial domain, RFID localization has many potential uses in warehousing and in supply chain management. In the healthcare domain, there are important uses for tracking/locating patients, medications, and instruments in hospitals, as well as tracking people with special needs. In the public safety and military domains, precise RFID localization can be utilized to assist firefighters and military personnel in accomplishing their missions [6] [7].

The emerging RFID technologies are based on a number of wireless inter-operability standards, which specify varied requirements as to power level, modulation technique, occupied bandwidth, frequency of operation, and so forth. They also utilize a variety of PHY and MAC layer techniques. Examples of these technologies include WiFi RFID (IEEE 802.11 a,b,g,n WLAN), UWB RFID (IEEE 802.15.3 WPAN), and ZigBee RFID (IEEE 802.15.4 WPAN).

Localization techniques in general utilize metrics of the received radio signals. Traditional metrics for localization applications are angle of arrival (AOA), received signal strength (RSS), time of arrival (TOA), and time difference of arrival (TDOA). Respective channel models targeting the behavior of each of these metrics for telecommunication purposes have been developed and are available in the literature. However, models for indoor localization applications must account for the effects of harsh indoor wireless channel behavior on the characteristics of the metrics at the receiving side, characteristics that affect indoor *localization* applications in ways that are very different from how they affect indoor *telecommunication* applications.

For example, the existing narrowband indoor radio channel models designed for telecommunication applications [8] can be used to analyze the RSS for localization applications, and the emerging 3D channel models developed for smart antenna applications [9] might be used for modeling the AOA for RFID localization. However, the available wideband indoor multipath channel measurement data and models presented in [2] are not suitable for analyzing the behavior of TOA and TDOA for localization applications.

The availability of several different RFID localization techniques employing different signal metrics provides the designer with various options to suit the specific application [10]. However, the ability to empirically evaluate the performance of a chosen RFID localization system configuration is important to successful design and deployment.

In this chapter we will introduce a novel real-time testbed for comparative performance evaluation of RFID localization systems, which can assist the system designer in evaluating the performance in a specific indoor environment. In the remainder of the chapter first we will provide an overview of indoor localization systems, their metrics, and their respective accuracies. Then we will provide a detailed description of the testbed and some performance evaluation results with emphasis on WiFi RFID localization in an indoor environment.

6.2   Fundamentals of RFID localization

The basic concept of any positioning system is illustrated in Fig. 6.1. The location-sensing devices measure the distance-related metrics, between the RFID device and a fixed reference point (RP), such as TOA, AOA, RSS, or TDOA. The positioning algorithm processes the reported metrics to estimate the location coordinates of the RFID tag. The display system exhibits the location of the mobile terminal to the user. The accuracy of the location estimation is a function of the accuracy of the location metrics and the complexity of the positioning

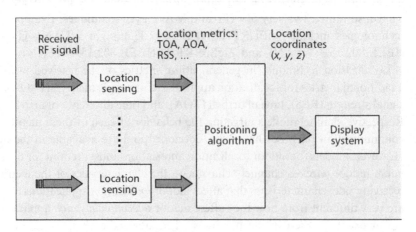

Fig. 6.1   A basic block diagram of a positioning system.

algorithm. There are two fundamental approaches to providing the infrastructure for an indoor positioning system. The first approach utilizes existing infrastructure already implemented for WLAN communication and adds a signaling system for RFID tags. The second approach requires implementing a specific infrastructure for the indoor positioning application.

This section is organized into three subsections. In the first two subsections, we discuss, respectively, AOA and TOA metrics, which are widely used in centralized localization. In the last subsection we discuss the RSS metric, which can also be used along with pattern-recognition techniques to estimate the location of the user [11].

Angle of arrival (AOA)

As the name implies, AOA gives an indication of the direction of the received signal. In order to estimate the AOA, the RFID tag and/or RPs must be equipped with a special antenna structure, specifically, an antenna array. In practice, severe multipath conditions in an indoor environment can lead to erroneous estimation of the AOA. Thus, relying on the AOA techniques alone in those environments would result in location estimates of low accuracy [12]. However, AOA estimation can be employed under adverse channel conditions to enhance the estimation of the TOA, as discussed in the next section.

Time of arrival (TOA)

In TOA-based localization, the TOA of the first detected peak (FDP) of the received signal is used to determine the time of flight between the transmitter and the receiver and consequently the distance between the transmitter and the receiver [12]. The channel impulse response (CIR) is generated by ray-tracing (RT) software for different locations of transmitter and receiver. The ideal CIR is usually referred to as the *infinite-bandwidth channel profile* since the receiver could theoretically acquire every detectable path. In practice, however, the channel bandwidth is limited. Filtering the CIR with a limited-bandwidth filter results in paths with pulse shapes. It can be shown that the sufficient bandwidth for accurate RFID localization based on the TOA metric is around 200 MHz [13].

Received signal strength (RSS)

The RSS is a simple metric that can be measured easily and it is measured and reported by most wireless devices. For example, the MAC layer of the IEEE 802.11 WLAN standard provides RSS information from all active access points (APs) in a quasi-periodic beacon signal that can be used as a metric for localization [12]. The RSS can be used in two ways for localization purposes, and we discuss these next.

Since the RSS decays linearly with the log-distance between the transmitter and receiver, it is possible to map an observed RSS value to a distance from a transmitter and consequently determine the user's location by triangulation using distances from three or more APs. That is, quantitatively,

$$RSS_d = 10 \log_{10} P_r = 10 \log_{10} P_t - 10a \log_{10} d + X \tag{6.1}$$

where a is the distance–power gradient, X is the shadow fading (a lognormal-distributed random variable), P_r is the received power, and P_t is the transmitted power. While simple, this method yields a highly inaccurate estimate of distance in indoor areas, since the instantaneous RSS inside a building varies over time, even at a fixed location; this is largely due to shadow fading and multipath fading [12]. If, on the other hand, we know what RSS value to expect at a given point in an indoor area, then we can estimate the location as the point where the expected RSS values approximate the observed RSS values most closely. This is the essence of the pattern-recognition approach to position estimation.

In the next section we present a framework for performance evaluation of indoor positioning systems and describe the structure of the testbed.

Description of the testbed

In this section we first describe the architecture of the testbed for evaluating the performance of indoor positioning systems and then provide the details of each block in the system. The performance of a positioning system is usually evaluated by collecting and recording massive amounts of measurement data. However, this approach is burdensome and time-consuming since the calibration step in the operation of a positioning system is typically extensive. In addition, this approach is not repeatable for different positioning systems at the same site, since the indoor propagation environment can change rapidly over time. Thus, the performance of a system might drastically degrade or improve from one data-collection interval to another. The testbed we describe here provides a repeatable environment in which to evaluate the performance of any positioning system and compare the levels of performance achievable with alternative systems.

General architecture of the testbed

The core of the testbed is a PROPSim C8 real-time channel simulator, developed by Elektrobit Corporation[1] which emulates the wireless channel between the transmitter and receiver. The other blocks of the testbed are the Ekahau positioning system, which utilizes WiFi RFID localization techniques, and RT software. Figure 6.2 shows the actual testbed in which the RT is merged with the PROPSim software, both implemented on a PC, while the Ekahau system and display module are installed on laptops.

Figure 6.3 shows the block diagram of the testbed. The PROPSim equipment is connected to the RPs in a shielded environment. The channel models are fed into

---

[1] http://www.elektrobit.ch.

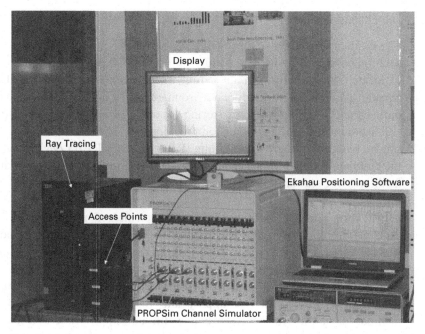

Fig. 6.2 Details of the elements of the testbed.

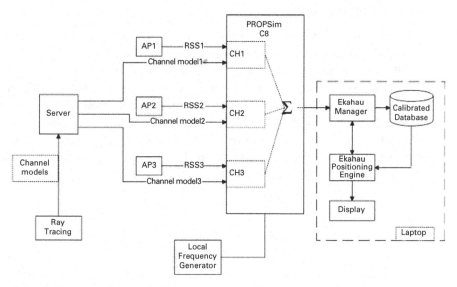

Fig. 6.3 A block diagram of the testbed.

PROPSim from RT. In the configuration shown here, we utilize three access points as RPs, and they are connected to the input ports of PROPSim. The output of PROPSim, which carries the emulated signals over a wireless channel, is connected to Ekahau through a network card.

The Ekahau Positioning Engine

The Ekahau Positioning Engine (EPE) is a software product based on an RTLS solution developed by Ekahau.[2] It uses an RSS-based pattern-recognition method to locate an object by communicating with the RPs and WiFi RFID devices already available in the indoor environment. The EPE constructs a database during the calibration phase by storing the signal strength measurements at several locations (i.e. training points) in the floorplan. The calibration measurements can be based on either access points (RPs) or other RFID devices available in the building. Ekahau Manager provides the user interface for the EPE; it initially facilitates the calibration process and is subsequently used for localization purposes.

PROPSim C8 is a real-time hardware-implemented radio channel emulator developed by Elektrobit that can emulate up to eight independent or correlative wideband radio channels at a time [14]. It uses a tapped-delay line model to simulate the channel between the input and output of each PROPSim channel. The proposed bandwidth of 70 MHz for a single independent channel of PROPSim makes the device appropriate for RSS-based RFID localization methods, since they consume less bandwidth, and appropriate to a lesser degree for TOA-based localization methods, which require greater bandwidth.

PlaceTool ray tracing

PlaceTool is a CWINS-developed software package that uses a 2D ray-tracing algorithm to model radio propagation in a typical indoor environment used for WLAN applications [15]. This software models all probable paths between the transmitter and the receiver on the basis of reflections from and transmissions through walls and barriers using information provided in the floorplan. The results include TOA, magnitude, phase, and direction-of-arrival information for each path. To provide an example, the following section describes the scenarios and details of experiments used to analyze the effects of different parameters on one indoor localization system.

6.3   Performance evaluation

To analyze the effects of different parameters on the performance of the example localization system, we initiated different scenarios, which will be discussed in the next subsections. The scenarios were intended to simulate the third floor of the Atwater Kent Laboratory building at Worcester Polytechnic Institute.

---

[2] http://www.ekahau.com.

Experimental detail

The first experiment was intended to analyze the effect of the number of RFID RPs (WiFi access points) and number of training points. We increased the number of RPs from one to three, and also raised the number of training points from 4 to 27. In the total of nine scenarios, we evaluated the performance of Ekahau; the results will be discussed in the next subsection. We conducted another experiment to analyze the effect of deployment strategy on the performance of the sample positioning system. We created three different network topology scenarios, each using three RPs, to evaluate the performance of Ekahau [16].

Results and analysis

To analyze the effects of different parameters on a positioning system, we begin by assessing the effects of the number of RPs and the number of training points. The channel responses are provided by RT and post-processed in Matlab®. After simulating the wireless channel using PROPSim for each location of the transmitter and receiver, the power metrics from the RPs are recorded by Ekahau for calibration of the positioning engine. In the next step, after constructing the pattern-recognition database using Ekahau, for a random location of the receiver, the process is repeated to record the power measurements of the respective receiver location and compare them with the database for pattern recognition. In this example, the grid network for location estimation consists of 66 points. The statistics of the scenarios are presented in the next subsection.

Numbers of RPs and training points

Figure 6.4(a) shows the average value of ranging error observed in the scenarios described above. As can be observed in Fig. 6.4(a), increasing the number of training points enhances the accuracy of the positioning engine. Using 27 training points for the sample indoor environment seems to be sufficient for localization exploiting pattern recognition. In addition, it can be seen that, in general, increasing the number of RPs enhances the performance of the localization system as well. The only observed inconsistency is for the case in which only one RP is used with few training points.

Figure 6.4(b) summarizes the effect of the number of RPs and the number of training points on the standard deviation of the ranging error in the experiments of this example. As can be seen, for the scenarios with one RP or two RPs, increasing the number of training points increases the standard deviation of ranging error. This is contrary to our expectation that adding training points should result in decreasing ranging error values. One possible explanation is that with fewer than three RPs Ekahau is unable to build an accurate database, which leads to substantial ranging errors.

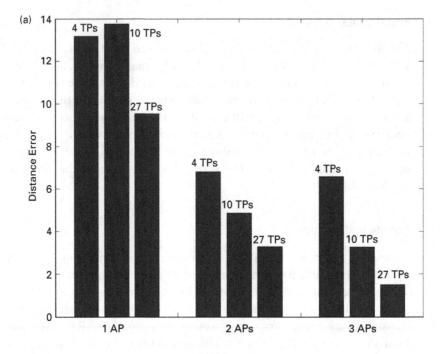

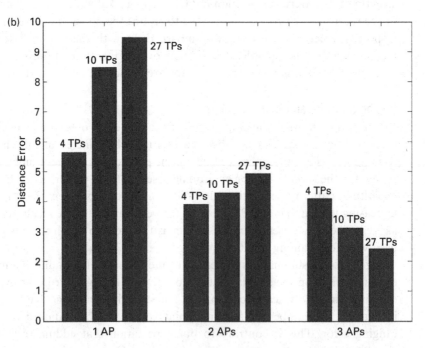

Fig. 6.4   Statistics of ranging error for different scenarios: (a) mean and (b) standard deviation of ranging error.

As expected, the number of RPs and number of training points are both significant in determining the accuracy of the positioning system. Increasing the number of RPs can greatly improve the accuracy of the positioning system. The same concept applies to the number of training points used to calibrate the positioning engine.

The sample cumulative distributive function (CDF) for different experiments is shown in Fig. 6.5. It can be seen that utilizing three RPs and enough (i.e. 20) training points means that the probability of having ranging errors smaller than 5 meters is greater than 90 percent, whereas using only one RP results in a ranging error of about 20 meters.

To analyze the effect of deployment topology on the performance of the sample localization system we created three different scenarios. Each scenario included three RPs, Ekahau was calibrated with 25 training points, and the grid network for location estimation consisted of 100 points. These three configurations for location of RPs are shown in Fig. 6.6.

The deployment of RPs was intended to cover the entire building for telecommunication purposes. The CDF of ranging error observed in each experiment is shown in Fig. 6.7(a). It can be observed that increasing the separation distance of the RPs within the coverage of each RP will result in an improved accuracy of the localization system. Figure 6.7(b) summarizes the results of the deployment experiment for different scenarios.

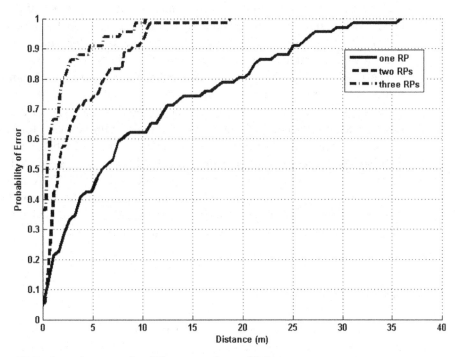

Fig. 6.5   CDF of ranging error for different numbers of RPs.

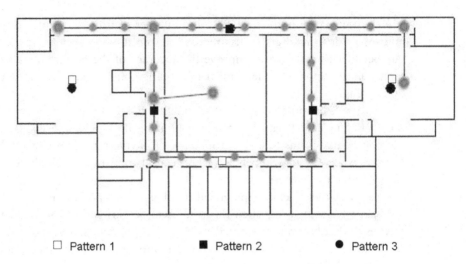

□ Pattern 1   ■ Pattern 2   ● Pattern 3

Fig. 6.6   Different strategies for deployment of the RPs.

6.4   Summary and conclusions

There are several factors that can affect the accuracy of location estimation in RSS-based localization systems. It is apparent that increasing the number of training points enhances the accuracy of estimation. With respect to reference points (RPs), the number of RPs installed has a strong influence on the accuracy of the localization system, as does the positioning of RPs within the building. The repeatability of experiments makes the testbed particularly valuable for performance evaluation of alternative localization systems incorporating WiFi RFID tag devices.

The ability to adapt the testbed described here for use with other distance-related metrics of the received radio signal is also very valuable. In the experiments described above we utilized RSS-based localization for purposes of example and investigated the effects of different environmental parameters vital for RSS-based localization. Simulating the AOA of the received signal will enable comparisons of RSS-based localization systems with AOA-based localization systems. Finally, as more accurate RFID localization systems migrate to the use of TOA as their radio metric, it would be valuable to extend the use of the real-time testbed to TOA-based systems. Minor modifications are needed to adapt the testbed presented here to the challenges associated with TOA-based localization techniques.

6.5   Acknowledgments

The authors would like to thank Dr. Allen H. Levesque for helpful discussions and detailed editorial comments for preparation of this chapter and Muhammad A. Assad for a review and comments.

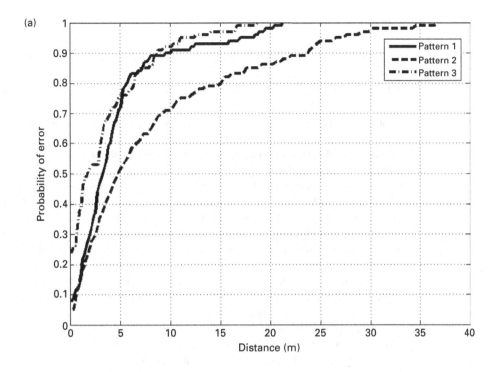

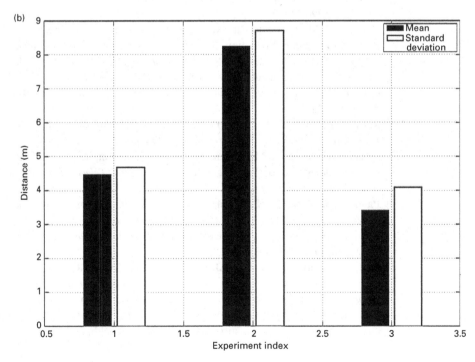

Fig. 6.7 Comparison of the ranging error for different deployment strategies. (a) Comparison of CDF of the ranging error. (b) Comparison of mean and standard deviation of the ranging error.

6.6   References

[1] **Kaplan, E. D.**, *Understanding GPS: Principles and Applications* (Artech House, Boston, MA, 1996).

[2] **Pahlavan, K.**, and **Levesque, A. H.**, *Wireless Information Networks*, 2nd edn. (John Wiley & Sons, New York, 2005).

[3] **Jules, A.**, "RFID Security and Privacy: A Research Survey," *IEEE Journal on Selected Areas in Communications*, **24**:381–394 (2006).

[4] **Phillips, T., Karygiannis, T.**, and **Kuhn, R.**, "Security Standards for RFID Market," *IEEE Security and Privacy Magazine*, 3:85–89 (2005).

[5] **International Organization for Standardization (ISO)**, *Information Technology – Real Time Locating Systems (RTLS) – Part 1: Application Program Interface* (2006) (http://www.iso.org/iso/en/CatalogueDetailPage.Catalogue Detail?CSNUMBER=38840&scopelist=PROGRAMME).

[6] **Sayed, A. H., Tarighat, A.**, and **Khajehnouri, N.**, "Network-based Wireless Location," *IEEE Signal Processing Magazine*, 22(4):24–40 (2005).

[7] **McKelvin, M. L., Williams, M. L.**, and **Berry, N. M.**, "Integrated Radio Frequency Identification and Wireless Sensor Network Architecture for Automated Inventory Management and Tracking Applications," in *Diversity in Computing Conference, TAPIA*, pp. 44–47 (Association for Computing Machinery, IEEE Computer Society, and Computer Research Association, Albuquerque, NM, 2005).

[8] **Pahlavan, K., Krishnamurthy, P.**, and **Beneat, J.**, "Wideband Radio Propagation Modeling for Indoor Geolocation Applications," *IEEE Communication Magazine*, 36:60–65 (1998).

[9] **Hassan-Ali, M.**, and **Pahlavan, K.**, "A New Statistical Model for Site-specific Indoor Radio Propagation Prediction Based on Geometric Optics and Geometric Probability," *IEEE Transactions on Wireless Communications*, 1:112–124 (2002).

[10] **National Institute of Standards and Technology (NIST)**, *RFID-assisted Localization and Communication for First Responders* (2006) (http://www.antd.nist.gov/wctg/RFID/RFIDassist.htm).

[11] **Roos, T., Myllymaki, P.**, and **Tirri, H.**, "A Statistical Modeling Approach to Location Estimation," *IEEE Transactions on Mobile Computing*, 1:59–69 (2002).

[12] **Kanaan, M., Heidari, M., Akgul, F. O.**, and **Pahlavan, K.**, "Technical Aspects of Localization in Indoor Wireless Networks," *Bechtel Telecommunications Technical Journal*, 5(1):47–58 (2006).

[13] **Pahlavan, K., Heidari, M., Akgul, F. O.**, and **Hatami, A.**, "Indoor Geolocation in the Absence of Direct Path," *IEEE Wireless Communications*, 13(6):50–58 (2006).

[14] **Kolu, J.**, and **Jamsa, T.**, *A Verification Platform for Cellular Geolocation Systems* (Elektrobit Oy, Oulunsolo, 2002).

[15] **Holt, T., Pahlavan, K.**, and **Lee, J. F.**, "A Graphical Indoor Radio Channel Simulator Using 2D Ray Tracing," in *IEEE Proceedings of Personal, Indoor, and Mobile Radio Communications*, pp. 411–416 (IEEE, Piscataway, NJ, 1992).

[16] **Unbehaun, M.**, *On the Deployment of Unlicensed Wireless Infrastructure*, unpublished Ph.D. dissertation, Royal Institute of Technology, Stockholm (2002).

7   Modeling supply chain network traffic

John R. Williams, Abel Sanchez, Paul Hofmann, Tao Lin,
Michael Lipton, and Krish Mantripragada

7.1   Introduction and motivation

In the future, when the Internet of Things becomes reality, serialized data (typically
RFID and/or barcode, based on EPCglobal, DOD/UID, and other standards) can
potentially be stored in millions of data repositories worldwide. In fact, large data
volumes of serialized information may be coming soon, as the global healthcare
industry moves towards deploying anti-counterfeiting standards as soon as 2009.
Such data will be sent to enterprise applications through the EPC network infrastructure. The data volume, message volume, communication, and applications with
EPC network infrastructure will raise challenges to the scalability, security, extensibility, and communication of current IT infrastructure. Several architectures for
EPC network infrastructure have been proposed. So far, most pilots have focused on
the physical aspects of tag readings within a small network of companies. The lack of
data quantifying the expected behavior of network message traffic within the future
EPC network infrastructure is one of the obstacles inhibiting industry from moving
to the next level. This chapter presents a simulator aimed at quantifying the message
flows within various EPC network architectures in order to provide guidance for
designing the architecture of a scalable and secure network.

RFID/EPC technology enables the tracking of physical objects through their
lifecycles without direct human involvement. Through the wide range of initiatives,
such as the one with retail giants (Wal-Mart and Target), and those with the Food
and Drug Administration (FDA), numerous state boards of pharmacy, aerospace
companies (Airbus and Boeing), and the Department of Defense (DoD),[1] RFID/
EPC/UID has demonstrated its great value for business operation automation.
Taking an airplane part as an example, it has the potential to be in any part of the
world. Therefore, the data for tracking this part can be recorded from any location.
Considering the diversity of organizations potentially dealing with this part
(manufacturer, airline, maintenance and repair companies), there could be thousands
of data repositories that might record the information related to this part.

---

[1] See http://www.acq.osd.mil/dpap/UID/.

RFID Technology and Applications, eds. Stephen B. Miles, Sanjay E. Sharma, and John R. Williams.
Published by Cambridge University Press. © Cambridge University Press 2008.

The data stored in the data repositories and also on RFID tags with the new IT infrastructure together form the EPC network infrastructure for the Internet of Things. Considering data operations with an airplane part, the number of data repositories, the data volume, the number of messages through the network, the business operations, and business applications involved are potentially far beyond the capacity of today's IT infrastructure.

With the joint effort by academia and industry, RFID/EPC technology has made great progress in the past few years. Up to now, most pilots have focused on the physical aspects of tag readings within a small network of companies. No quantified data has been collected for the potential EPC network infrastructure. Several architectures for the EPC network infrastructure have been proposed. However, due to the lack of a mechanism for evaluating these architectures with quantified data, no common criteria can be researched.

This chapter presents a supply chain simulator in order to obtain the quantified data of the message flow in the future EPC network infrastructure. The design and development of such a simulator is a complicated task because a number of dimensions needs to be considered, such as scalability, extensibility, security, privacy, communication frequency, and in-time response.

The rest of this chapter is organized as follows. Section 7.2 analyzes the requirements of the simulator through a pharmaceutical use case. Section 7.3 presents the software architecture for this simulator environment. Section 7.4 discusses a few implementation issues and gives some initial results. Section 7.5 concludes this chapter.

7.2   Requirements

The pharmaceutical supply chain

Counterfeit and compromised drugs are increasingly making their way into the public healthcare system and are considered a threat to public health by the FDA [1]. Counterfeit pharmaceuticals are a $32 billion dollar industry representing 10 percent of the global market, according to the FDA. The recent increase in number of patients in the USA receiving fake or diluted drugs is focusing more attention on the need for drug authenticity. In 2003, 18 million tablets of the cholesterol-level-lowering drug Lipitor, the world's best-selling prescription drug in 2004, were recalled by Pfizer in the United States after fake pills had been found in pharmacies [2]. In 2004, the FDA reported 58 counterfeit drug cases, a 10-fold increase since 2000 [3].

Supply chains consist of several kinds of enterprises, such as manufacturers, transportation companies, wholesalers, and retailers. The pharmaceutical supply chain is one of the most complex supply chains and can have as many as a dozen or more enterprises between the manufacturer and the customer. The growth of counterfeiting has recently led a number of states in the United States of America to consider passing laws requiring that the pedigree of every salable unit of drugs be tracked. To date, over 30 states have proposed or passed pedigree legislation. In the case of California a digital document tracking each salable unit with a unique identifier must be initiated by the manufacturer, transmitted downstream step by step to the dispensing pharmacy, and appended to by each enterprise along the way, with digitally signed details of every shipping and receiving event. Several proposals for how this digital document should be produced are being considered. The FDA Counterfeit Drug Task Force has recommended "a combination of rapidly improving 'track and trace' technologies and product authentication technologies" to protect the pharmaceutical drug supply [4].

Four feature characteristics are paramount to widespread adoption and impact.

- **Automation** – the high volume and high variance of pharmaceutical packages makes it impractical for supply chain participants to economically authenticate packaging manually. Therefore, there is a need for authentication that is automated – needing little to no human involvement or interpretation to authenticate the packaging. Automated strong authentication requires electronic acquisition of information from product packages in mass and without special handling.
- **Security** – in order to have high confidence that the product is authentic, the expected features of the package, namely physical, electronic or some combination of the two, must be difficult or economically impractical to duplicate and simulate.
- **Privacy** – concerns for consumer privacy must be respected.
- **Timeliness of response** – since companies cannot ship or use product until pedigrees have been received and authenticated, timely response for all pedigree-related transactions is required.

Options for ePedigree

In varying degrees, all of the state and federal legislative initiatives require a document to be passed along the supply chain along with the physical product. Many of these states, such as California, require an electronic document tied to unique serial numbers. The ePedigree standard recently ratified by EPCglobal Inc. provides a ratified XML schema for such a document (see Fig. 7.1).

Thus, when the manufacturer makes a shipment of, say, N items, a pedigree document listing the manufacturer, the date of shipment, the type of product, and the EPC codes of all the items will accompany the shipment. A hash of this document is then computed and signed with the manufacturer's private key (using the public key infrastructure). Upon receiving the goods the wholesaler will add to the ePedigree details of the receipt and will again hash and sign the document. The wholesaler will then add further to the document upon shipment. This process is repeated at every receipt and shipment event until the goods reach the dispensing pharmacy. At each level, the signed inbound ePedigree documents must be embedded into the outbound document, creating a complex nested document.

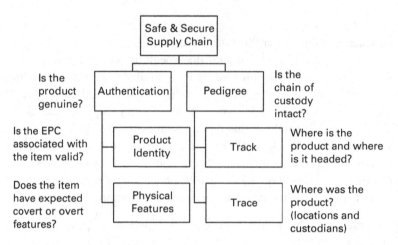

Fig. 7.1   The base reference model for a secure supply chain (EPCGlobal Inc. e-Pedigree).

One disadvantage of this system is that downstream customers may gain knowledge of upstream enterprises' business practices. For example, if the manufacturer produces only one ePedigree document for, say, 500 items, then everyone downstream can see that the manufacturer shipped this quantity of goods. To obviate this issue, manufacturers may choose to produce a separate ePedigree document for every item (or case) shipped.

There is concern in the industry that the size and number of ePedigree documents will be large, resulting in a problem as the system scales up.

The registry concept

Several alternatives to the document-passing scheme have been proposed. One alternative is that ePedigree "fragments" remain with the "owner" and that the "fragments" be assembled "on the fly" only if required by some authority. Thus, instead of passing on the actual ePedigree document, a link to this "fragment" would be either passed along the supply chain or possibly passed to some third-party registry. Within this concept there are numerous variations, with more or less information stored in the registry, implying more or less effort to assemble the pedigree when required.

A supply chain network simulator

In order to assess the implications of various approaches to granularity, security, and alternative pedigree models, a simulator is being developed. The simulator is composed of N supply chain tiers, such as manufacturer, wholesaler, and retailer, where each tier may have an arbitrary number of facilities. Each facility is modeled as a state machine running in its own thread of execution.

Just like the links in a metal chain, the members of a supply chain may have business relationships only with their immediate neighbors. They may, or might not, know about more distant members of the chain and even if they are aware of their existence they might not have a business relationship with them.

The supply chain functions by executing business events between trading partners. One party initiates an event by sending a message to the other party, such as a purchase order (PO) message.

The state of a facility is determined by the number of POs it has pending and how much stock it has accumulated in its warehouse. The simulation is driven by POs that are submitted "upstream" by the retail tier. Goods are manufactured in response to purchase orders and are shipped "downstream." Initial results show that the simulator is capable of modeling 100,000 facilities and 100 million items of product being injected into the system per day. The load on the registry can vary by a factor of over 1,000 from peak to average load, with around 200 messages per second being the peak load for a flow of 1 million per day.

7.3   Software architecture

The system is designed to run on a single machine in a massively threaded environment (Fig. 7.2). Each facility is totally independent of all other facilities and interacts by receiving messages on "ports." Facilities are known by their "FacilityID" and the scheduler maintains a registry of "endpoints" that can be either references to facility objects on the same machine or web service endpoints on a different machine.

The physical supply chain is organized into tiers, such as Manufacturing Tier, Tier 1 Wholesaler, Tier 2 Wholesaler, and so on with the Retailer Tier being the

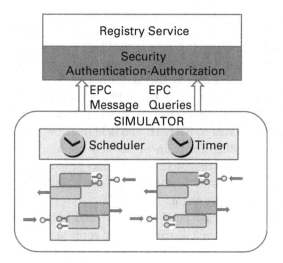

Fig. 7.2   Simulator layers with scheduler and registry services.

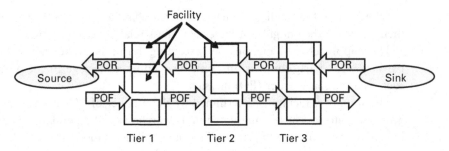

Fig. 7.3   The supply chain organized into tiers and facilities.

final tier (Fig. 7.3). The physical goods flow downstream from the manufacturer to the retailer. Purchase orders are propagated upstream until a facility is able to satisfy them.

There are two special facilities in our model that are not in the physical supply chain, namely the source and the sink. These are named from the perspective of the production of physical goods because the source acts as the universal producer of goods. The sink is the universal purchaser of goods and also simulates purchasers' demands by issuing PO requests into the system.

Both the source and the sink can be used to control the flow of goods through the system. For example, the flow of goods through the system can be controlled by the sink issuing PO requests to the Retail Tier. For example, if the sink issues requests for 1 million items per day then, once the purchase orders have reached the manufacturers and enough time has elapsed to reach steady state, the average flow of physical goods through any tier will be 1 million items per day. This can easily be seen since only the universal source can manufacture goods and there is no loss of goods in the system, and the capacity of the warehouses is finite.

Purchase order requests

When a PO request message is posted to a facility the facility checks its warehouse for the items required (Fig. 7.4). If the warehouse can fulfill the order then the items are sent to "shipping," where they are stored until shipped. If the warehouse cannot fulfill the order then the order is sent to the PO Consolidation Store and a copy is sent to the PO Pending Fulfillment Store. The items sent to the PO Consolidation Store are aggregated by SKU ID and held there until a trigger from the scheduler initiates the sending upstream of the new consolidated POs. The facility therefore generates new PO requests and these may be issued in "bursts" at various times of the day. These bursts do not generate event traffic to the registry but do add variability to the supply chain.

The facilities are able to aggregate purchase orders and hold them for some period of time. In a real supply chain a distributor may apply various strategies to manage their inventory, including pre-ordering items on the basis of past history. The simulator is able to accommodate such strategies.

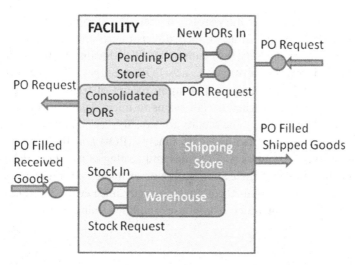

Fig. 7.4   The facility state machine showing incoming and outgoing events.

Purchase order fulfillment

When physical goods representing a filled inbound PO arrive at a facility they are immediately moved to the warehouse. When new stock arrives in the warehouse the store of outbound POs pending fulfillment is scanned to see whether any can now be filled. If an outbound PO request can now be filled the items are removed from the warehouse and sent to shipping, where they await a shipping event trigger from the scheduler.

7.4   Implementation

Modeling facilities as state machines

Each facility is represented as a state machine running in its own thread of execution. The state of the facility is modified by two kinds of events, namely PO request (POR) messages that represent purchase orders and PO filled (POF) messages that accompany goods being received. These messages are received on two external message ports, one for each kind of message. The state of the facility is represented by the following.

1. The number and type of goods stored in the warehouse as a result of goods received,
2. The Unfulfilled POR Store that keeps track of PORs that have been sent upstream but have not yet been filled, i.e. goods are not yet available in the warehouse to fill these orders.

There are also two temporary stores, namely the Shipping Store, where goods are accumulated before being shipped downstream, and the Consolidated PO Store, where incoming POs that cannot be filled locally by the warehouse are aggregated and then sent upstream as new POR messages. These temporary stores are "emptied" and the messages fired upon receipt of trigger messages from the global "scheduler." The scheduler allows us to inject delays into the facility that represent the time taken by the facility to, say, ship goods from the warehouse.

These two stores, one for digital documents (POs) and the other for physical goods (warehouse) allow us to add rules and strategies into how POs are aggregated and the timing of their submittal. For example, we might consolidate all POs for one day and submit them upstream only once every 24 hours. Similarly, we can vary the way in which warehouses fulfill incoming POs.

The Scheduler

The scheduler keeps track of all the facilities. It also provides a number of control parameters that determine the rate at which manufactured goods are injected into the supply chain and the rate at which POs are injected by the retailers in response to goods being sold. It also provides "timers" that send trigger events to the facilities that control when goods are shipped downstream and when POs are sent upstream.

The messages for POR and POF are inherited from the base POMessage class shown below. Each message contains a PO and the address of the sender of the message. The scheduler is responsible for translating FacilityIDs to actual endpoints to which the message is delivered. These endpoints are called "ports" and are the building blocks of our simulator (Fig. 7.1):

```
public class PO
  {
    public FacilityID nextDestination;
    public int skuID;
    public int numItems;
  }
    public class POMessage
  {
    public PO body;
    public FacilityID senderFacilityID;
  }
```

The port abstraction

One novel aspect of the simulator is that it is built upon the Microsoft Coordination and Concurrency Runtime (CCR) [5]. The runtime provides support for multi-threaded programming. It provides an abstraction called a port (Fig. 7.5) that deals with messages of a single type. A port allows messages to be "posted" to it and there is a buffer that stores the messages. A multi-cast delegate can be

Port

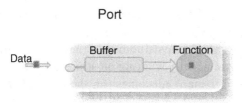

Fig. 7.5   The port abstraction, showing the buffer and the delegate function.

attached to the port and, when messages arrive at the port, they are passed to the delegate for processing. There is the concept of "activating" a port, which triggers the passing of the messages to the delegate function(s). The way in which messages are buffered can be controlled so that, for example, we can wait until N messages have arrived before passing them to the function.

A port with no messages in its buffer consumes around 175 bytes of memory, so it is possible to create around 7 million ports on a laptop. Since these "internal ports" correspond closely to sockets, it is quite easy to arrange for a port to pass on its messages to an external web service for processing. Thus, to simulate the messages to the registry, we can easily push messages representing shipment or receipt of goods to or from a facility to a registry web service on a remote machine.

The CCR arranges for every port to run in its own thread of execution. Thus, once POR or POF messages have been injected into the simulator, the facilities respond to these messages independently (as autonomous state machines).

The registry

The simulator is capable of simulating the message traffic both between facilities and also to a third-party registry. There are two kinds of message traffic to the registry, namely "I've touched EPC X" messages and query messages of the kind "Who has seen EPC X?" The first kind of message traffic results from shipments received (incoming POFs) and from shipments shipped (outgoing POFs). This traffic places into the registry notification events for each EPC code involved. A typical message might contain the following: EventType, EPC, shipper, Date-Time, receiver, PedigreeHash, PedigreeURI.

According to the EPCIS 1.0 specification this traffic is likely to aggregate together all EPCs read at one time (we note that there is as yet no specification for these messages). It is likely that these messages will conform to either the EPCIS 1.0 specification or something quite similar. (Appendix 7.1 shows an EPCIS 1.0 conformant ship order event.)

The registry will probably respond to such messages with a simple acknowledgment, and therefore the incoming messages can be buffered by the registry to smooth any peaks in the message traffic.

The second kind of message traffic results from queries, such as pedigree queries, which require a more elaborate response and are likely to be time-sensitive. Thus, these queries should be answered in a "timely" manner by the registry. The query response time will depend on the volume of data to be searched (EPC data must be stored for several years) and therefore partitioning of the registry data may be critical. MIT and SAP Labs are currently working on strategies for partitioning in-memory databases spread over many machines (on the basis of the Google model).

Security and authorization

One element affecting the latency of the registry responses is how authentication and authorization will be achieved. If there is a single registry and all supply chain participants have pre-established security credentials then there are standard methods for dealing with both authentication and authorization. Authentication (who is making the query) can be established by using X.509 certificates as part of a public-key infrastructure (PKI). Authorization requires the registry to answer the question "Does this person have permission to retrieve data related to this EPC?" This is a more complex question because to answer it the registry must have knowledge of the kind of item to which the EPC is attached. For example, the EPC might be attached to a packet of cornflakes or to a cruise missile. In the latter case, even details about what companies are involved in the supply chain might be secret, let alone details about the present location of the missile. Thus, the registry must have "business rules" that depend on the type of item (its SKU) and on the roles associated with that item.

If there are multiple registries with multiple security boundaries (i.e. registries operated by different enterprises) then the problem becomes more complex, since standards for communicating queries between registries need to be established. This problem is currently being researched by EPCGlobal Inc., the Architectural Review Committee, and the Auto-ID Laboratories.

7.5   Simulator performance

MIT and SAP Labs are currently working on applying the simulator to analyze potential network traffic of the pharmaceutical supply chain. Inputs for this model consist of item throughput statistics and queries to the registry. With this model we expect to be able to develop envelopes for the capabilities necessary for the various kinds of registry.

The present simulator, running on a Dell Latitude 620, is able to represent around 1 million facilities, which consume around 1 GByte of memory. The CCR is very efficient in that only those facilities which are "actively" processing messages consume resources. The performance of the simulator is ultimately limited by the message traffic. Initial runs have simulated a volume of 1 million items per day with the simulator running in real time.

7.6   References

[1] **FDA**, *Combating Counterfeit Drugs: A Report of The Food and Drug Administration Annual Update*, p. 1 (2005) (http://www.fda.gov/bbs/topics/NEWS/2005/NEW01179. html).

[2] **CIO**, *Cracks in the Pharmaceutical Supply Chain*, p. 1 (2006) (http://www.cio.com/article/16565/Cracks_in_the_Pharmaceutical_Supply_Chain/1).

[3] Reference [1], p. 2.

[4] **FDA**, *Combating Counterfeit Drugs* (2004).

[5] **Chrysanthakopoulos, G.**, and **Sing, S.**, *An Asynchronous Messaging Library for C#* (http://research.microsoft.com/~tharris/scool/papers/sing.pdf).

7.7   Appendix

XML file of an EPCIS read event.

```
<?xml version=''1.0'' encoding=''UTF-8''
  standalone=''yes'' ?>
<epcis:EPCISDocument  xmlns:xsi=''http://www.w3.
  org/2001/ XMLSchema-instance''
xmlns:epcis=''urn:epcglobal:epcis:xsd:1''xmlns:
epcglobal=''urn:epcglobal:xsd:1''
xsi:schemaLocation=''urn:epcglobal:epcis:xsd:1
EPCglobal-epcis-1_0.xsd''
xmlns:hls=''http://schema.hls.com/extension''
creationDate=''2006-06-25T07:15:00Z'' schema
Version=''1.0''>
<EPCISBody>
<EventList>
<TransactionEvent>
<eventTime>2006-06-25T07:16:00Z</eventTime>
<bizTransactionList>
<bizTransaction type=''urn:epcglobal:fmcg:btt:
po''>urn:epcglobal:fmcg:bti:po:0614141073468.1
</bizTransaction>
<bizTransaction type=''urn:epcglobal:fmcg:btt:
bol''>urn:epcglobal:fmcg:bti:bol:0614141073468.A
</bizTransaction>
</bizTransactionList>
<epcList>
<epc>urn:epc:id:sgtin:0614141.107340.1</epc>
<epc>urn:epc:id:sgtin:0614141.107340.2</epc>
  </epcList>
<action>ADD</action>
<bizStep>urn:epcglobal:fmcg:bizstep:shipping</bizStep>
```

```
<disposition>urn:epcglobal:fmcg:disp:sellable_available
</disposition>
<readPoint>
<id>urn:epcglobal:fmcg:loc:0614141073468.RP-3</id>
  </readPoint>
<bizLocation>
<id>urn:epcglobal:fmcg:loc:0614141073468.3</id>
  </bizLocation>
  </TransactionEvent>
<TransactionEvent>
<eventTime>2006-06-25T07:17:00Z</eventTime>
<bizTransactionList>
<bizTransaction type=''urn:epcglobal:fmcg:btt:po''>
urn:epcglobal:fmcg:bti:po:0614141073468.2</bizTransaction>
<bizTransaction type=''urn:epcglobal:fmcg:btt:bol''>
urn:epcglobal:fmcg:bti:bol:0614141073468.B
  </bizTransaction>
  </bizTransactionList>
<epcList>
<epc>urn:epc:id:sgtin:0614141.107342.1</epc>
<epc>urn:epc:id:sgtin:0614141.107342.2</epc>  </epcList>
<action>ADD</action>
<bizStep>urn:epcglobal:fmcg:bizstep:shipping</bizStep>
<disposition>urn:epcglobal:fmcg:disp:sellable_available
</disposition>
  <readPoint>
<id>urn:epcglobal:fmcg:loc:0614141073468.RP-3</id>
  </readPoint>  <bizLocation>  <id>urn:epcglobal:fmcg:
loc:0614141073468.3</id>
</bizLocation>
  </TransactionEvent>  <TransactionEvent>
<eventTime>2006-06-25T07:18:00Z</eventTime>
<bizTransactionList>
<bizTransaction type=''urn:epcglobal:fmcg:btt:po''>
urn:epcglobal:fmcg:bti:po:0614141073468.3
  </bizTransaction>
<bizTransaction type=''urn:epcglobal:fmcg:btt:bol''>
urn:epcglobal:fmcg:bti:bol:0614141073468.C
  </bizTransaction>
  </bizTransactionList>
  <epcList>
<epc>urn:epc:id:sgtin:0614141.107344.1</epc>
<epc>urn:epc:id:sgtin:0614141.107344.2</epc>
  </epcList>
<action>ADD</action>
<bizStep>urn:epcglobal:fmcg:bizstep:shipping</bizStep>
<disposition>urn:epcglobal:fmcg:disp:sellable_available
```

```
  </disposition>
  <readPoint>
<id>urn:epcglobal:fmcg:loc:0614141073468.RP-3</id>
  </readPoint>
<bizLocation>
<id>urn:epcglobal:fmcg:loc:0614141073468.3</id>
  </bizLocation>
  </TransactionEvent>
<TransactionEvent>
  <eventTime>2006-06-25T07:19:00Z</eventTime>
<bizTransactionList>
<bizTransaction type=''urn:epcglobal:fmcg:btt:po''>
urn:epcglobal:fmcg:bti:po:0614141073468.4
  </bizTransaction>
<bizTransaction type=''urn:epcglobal:fmcg:btt:bol''>
urn:epcglobal:fmcg:bti:bol:0614141073468.D</bizTransaction>
  </bizTransactionList>
  <epcList>
<epc>urn:epc:id:sgtin:0614142.107346.1</epc>
<epc>urn:epc:id:sgtin:0614142.107346.2</epc>
  </epcList>
<action>ADD</action>
<bizStep>urn:epcglobal:fmcg:bizstep:shipping</bizStep>
<disposition>urn:epcglobal:fmcg:disp:sellable_available
  </disposition>
  <readPoint>
<id>urn:epcglobal:fmcg:loc:0614141073468.RP-3</id>
  </readPoint>
<bizLocation>
<id>urn:epcglobal:fmcg:loc:0614141073468.3</id>
  </bizLocation>
  </TransactionEvent>
<TransactionEvent>
t<eventTime>2006-06-25T07:20:00Z</eventTime>
<bizTransactionList>
<bizTransaction type=''urn:epcglobal:fmcg:btt:po''>
urn:epcglobal:fmcg:bti:po:0614141073468.5</bizTransaction>
<bizTransaction type=''urn:epcglobal:fmcg:btt:bol''>
urn:epcglobal:fmcg:bti:bol:0614141073468.E
  </bizTransaction>
  </bizTransactionList>
<epcList>
<epc>urn:epc:id:sgtin:0614142.107348.1</epc>
<epc>urn:epc:id:sgtin:0614142.107348.2</epc>
  </epcList>
<action>ADD</action>
<bizStep>urn:epcglobal:fmcg:bizstep:shipping</bizStep>
```

```
<disposition>urn:epcglobal:fmcg:disp:sellable_available
  </disposition>
  <readPoint>
<id>urn:epcglobal:fmcg:loc:0614141073468.RP-3</id>
  </readPoint>
<bizLocation>
<id>urn:epcglobal:fmcg:loc:0614141073468.3</id>
  </bizLocation>
  </TransactionEvent>
  </EventList>
  </EPCISBody></epcis:EPCISDocument>
```

8 Deployment considerations for active RFID systems

Gisele Bennett and Ralph Herkert

In transitioning from the analysis of RFID systems technology to actual deployment considerations of specific sites, the first area confronting project managers is the need to assess the range of other RF applications that may be operating at the same unlicensed frequencies.

A critical part of preparing for RFID system implementations is planning to support the various RFID applications infrastructures that may be implemented. Specifically, the opportunity to leverage a common IT infrastructure for supply chain, pedigree-tracking, and location-tracking RFID services promise higher return from investments in these technologies. One example is to consider leveraging a wireless data (WiFi) network as the back-haul for RFID readers. Another would be to analyze the costs and benefits of combining RTLS applications with the same WiFi infrastructure, rather than using stand-alone systems for both. At a minimum, a gap analysis is required, in order to assess systems operating across the unlicensed spectrum that might interfere with the performance of RFID systems, or vice versa.

8.1 Introduction

Wal-Mart and the Department of Defense's announcement in 2003 mandating their suppliers to implement RFID tagging for all goods supplied to them caused a flurry of panic and speculation on how best to implement this requirement. Other agencies, including the FDA and corporations, have followed suit, with varying timelines and deployment plans. The area of passive RFID is more mature than that of active RFID with respect to the level of standardization and test data. The majority of the logistics world has been implementing RFID for tracking using available technologies that only recently included RFID. Even in a shipping port, RFID extends beyond asset tracking to include access control, container security, container identification and location, and activity tracking. Since these announcements and the initial market flurry, passive RFID has had a

RFID Technology and Applications, eds. Stephen B. Miles, Sanjay E. Sharma, and John R. Williams. Published by Cambridge University Press. © Cambridge University Press 2008.

chance to mature, although it is still not at the level of a "plug and play" technology.

The area that is not as well defined is active RFID and especially active RFID with sensor integration. Active RFID requires, in most cases, more technology and policy considerations for implementers. Areas to consider in designing active RFID systems with integrated sensors include power (low power consumption and power management for extended life, battery recycling, emitted power for FCC compliance), sensor integration, secure data transmission, frequency interference and impact analysis, and general architecture design and implementation considerations. With added features, there are always challenges to managing the development and deployment costs and matching them with a business case. Business case examples illustrating this point are addressed in Ch. 9. This chapter focuses on engineering issues and methodologies for deploying active RFID systems, with an emphasis on systems, with integrated sensors.

8.2   Basics of the technologies

A typical RFID system consists of a tag, air interface between the tag and reader, an interrogator (or reader), and a back-end enterprise system or host. The tag is attached to an asset or object and, for active systems, contains a chip and memory, an antenna, maybe sensors, and a battery. The interrogator consists of the transmitting and receiving antennas for information transfer from a tag. The air interface is the means for data transfer between tag and reader and, for active systems, could include "tag-to-tag" data transfer. Although various guidelines exist for defining passive, semi-active, and active RFID systems, we can simplify the categories into a passive and active RFID architecture. A good overview of the differences between active and passive RFID is provided in a white paper compiled by Q.E.D. Systems [1] [2]. Both technologies utilize an interrogator and a tag. Passive RFID tags require an interrogator to energize the tag in order to receive the signal with an identifier for that tag. In Ch. 4 Marlin Mickle et al. provide an in-depth discussion of this operation and tag behaviors at these frequencies. An active RFID system utilizes the tag's own power source and, depending on the architecture, can transmit independently of interrogation. Advantages of active RFID include greater read range and an "on-board" power source for sensors. Active systems are more complex, since they generate additional information beyond the asset ID to be transferred to the reader. Features added to active RFID tags include greater memory for data storage, integrating sensors for condition monitoring (e.g. temperature, pressure, humidity, and shock), and chemical/biological sensor integration. With these additional features comes a cost, both in financial terms and in complexity. The resulting trade-offs that must be evaluated are discussed in the following section.

8.3   Technology and architectural considerations

Depending on the application, certain technical considerations may be applicable to specific use cases. For example, if tags are installed on assets that are remote and not accessed regularly, then power management is important in order to avoid the need to change batteries. A common element of all of the technology and architectural considerations is the importance of understanding the requirements for the desired application. What might be an ideal solution for one application could be unacceptable in another. Active tags have the luxury of additional memory on which one can store information about the asset, e.g. condition monitoring or location or data beyond the asset ID. All of these additions make the systems attractive but require consideration of other architectural and cost factors. If sensors are added to RFID tags, then a communication protocol for sensor data exchange must be considered. For a detailed review of a proposed RFID plus sensor protocol, see Ch. 5. Most active RFID systems considered in this chapter follow the ISO 18000 Part 7 standard.

Power

Power typically poses a challenge for active RFID tags. In most systems the readers are fixed or easily accessible mobile units that either have access to readily available power or are systems for which the batteries can easily be changed. On the other hand, active RFID tags are used in applications such as container monitoring and pallet tracking or could be used for some other remote condition tracking. Since these tags are not always accessible, changing the batteries is neither a practical nor a cost-effective approach to managing power consumption. In some cases milli-amp (mA) power consumption is a significant amount to contend with in an overall systems design. There are several approaches to managing power. If sensors are integrated then a microprocessor is probably available to collect the information. The tags, through this microprocessor, could be put into sleep mode according to a duty cycle that could be programmed to alleviate power consumption. In addition to the tag's own power, the communication module is typically a primary source of power consumption for long-distance applications, since the communication range is proportional to the power consumption. An introduction to tag optimization strategies is provided by Hao Min (Ch. 3) as a methodology for working to lower transmit/read ranges and minimizing power consumption.

Another consideration for power management is specifying the source, i.e. the battery. Although this too might appear to be a trivial task, the application and environmental constraints could complicate the choice. Consider the case of installing active tags on air cargo containers according to guidelines from the Federal Aviation Authority (FAA) and the Office of Hazardous Materials Safety on the transportation, storage, usage, and handling of batteries in air cargo. These regulations and guidelines are updated regularly and must be reviewed prior to air transport of the active RFID system. Expanding on FAA requirements, environmental operation ranges for the tags must also be considered because not all batteries will work under typical operating ranges from −40 °C to 70 °C in temperature.

As RFID technology matures, more options for power sources with flexibility in integration are coming onto the market, e.g. fuel cells on a chip and flexible (physically pliable) batteries that do not need recharging.

Sensor integration

Integrating sensors with active RFID systems provides a powerful advantage over passive RFID. However, be mindful that the integration of the sensors should be application-driven since the added cost of the tags must be justified by adding value to the application. When condition monitoring is desired, which could include monitoring of temperature, pressure, humidity, location, etc, then integrating sensors with active RFID can provide real-time asset location tracking and/or logging of sensor data en route or in storage. For how historical sensor information is used in "the cold chain" see Ch. 11. In addition to the added cost of active sensors, there are costs associated with power consumption. As noted in Section 8.2, it will be important to program the sensors to sample the environment at some appropriate interval rather than to perform continuous monitoring. With this additional data, other systems design considerations such as the amount of data to store and/or transmit and the frequency of data exchange will drive memory, storage, and transmission requirements. Other sensors that could be included in the overall system architecture, which would be separate from the tag, are the integration of barcode and/or contact memory readers that could transfer information from other sources to the active tag's memory for future remote access. The data content will drive the choice of security measures that should be implemented, as discussed in Ch. 7.

Networking protocols, layers, and mesh networking should be considered. Depending on the sophistication of the tag network, mesh networking can be used to enhance situational awareness of the location of assets relative to one another while reducing overall power consumption by the tags. Power reductions can be achieved by hopping information between active RFID tags that act as nodes for information transfer to a reader and eventually to the enterprise. Because an ad-hoc network is established, one byproduct of mesh networking is knowing where each tag (and associated asset) is located relative to others. If fixed tag locations are known, a map of the site can be drawn [3].

Frequency usage and power emitted

Currently there are eight different bands of frequencies used in RFID applications. Globally each country regulates its own frequency band for RFID [4]. The characteristics of each frequency band, low (100–500 kHz), intermediate (10–15 MHz), or high (850–950 MHz), have been discussed in Ch. 4. Radio frequency performance varies from short to long read ranges according to the propagation through materials and the power budget that is allowed at each frequency in specific jurisdictions.

Testing for compliance

There are various standards and regulations governing frequency emissions. These standards provide consistency for protocol interoperability and procedures for universal compliance testing. Some jurisdictions are international, national, industry-specific, and/or agency-specific as in the case of the Department of Defense (DoD). Agencies that impact RFID testing methodologies include but are not limited to the International Organization for Standardization (ISO), International Electrotechnical Commission (IEC), International Telecommunications Union (ITU), American National Standards Institute (ANSI), Uniform Code Council (UCC), Federal Information Processing Standards (FIPS), and Federal Communication Commission (FCC). The FCC regulation for transmitters operating from the LF to UHF bands is governed by Part 15 of Title 47 of the Code of Federal Regulations of the FCC rules.[1] This section will briefly discuss the various FCC approaches to operating in a frequency band.

FCC licenses are granted for allocation of a specific portion of the frequency spectrum to be used by the product. Licenses do not specifically pertain to individual products, but instead to the frequency spectrum the product uses to transmit data. Therefore, an individual product may not be granted a license, but the user would need to obtain an FCC license for transmitting on a particular frequency. The only way to avoid FCC licensing is to transmit in a license-free frequency band. Two main license-free frequency bands in the United States are 902–928 MHz and 2.4–2.4835 GHz.

FCC compliance refers mainly to the emissions of a product. FCC compliance simply means that the product meets all FCC specifications regarding spurious emissions. All products must go through testing to meet FCC compliance standards, but some products, such as PCs and printers, need nothing more than compliance testing. FCC certification is needed for RF products deemed intentional transmitters. FCC certification means that an FCC laboratory has tested the product, has officially certified the product, and has assigned an FCC identification number. Certification testing is much more rigorous than compliance testing and may be broken down into two categories: low-power products and high-power products. Low-power products are those that transmit in the milliwatt (mW) range, whereas high-power products have a transmission output power greater than two watts. Low-power testing generally takes less time than high-power testing and can usually be completed in a couple of days. The use of FCC pre-approved chipsets by interrogator OEMS may improve FCC certification testing performance.

The DoD compliance considers the following electromagnetic effects of RF propagation: (1) hazards of electromagnetic radiation to fuels (HERF),

---

[1] See http://www.fcc.gov/oet/info/rules/part15-2-16-06.pdf.

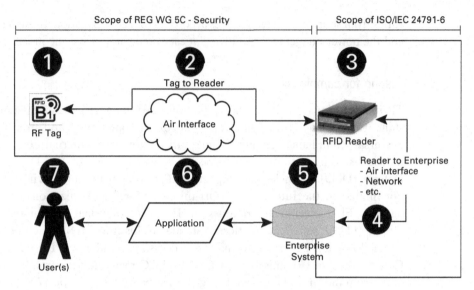

Fig. 8.1   RFID system architecture (from AIM RFID expert group (REG) on RFID – Guidelines on Data Access Security [AIM REG 314], with permission [7].

(2) hazards of electromagnetic radiation to personnel (HERP), and (3) hazards of electromagnetic radiation to ordnance (HERO).

Data security and privacy are the subjects of ongoing challenges and debates, especially in the defense industry. Privacy cannot occur without data security. Data security must be implemented throughout the entire chain of custody for data, starting with the tag to the reader, or from tag-to-tag RF transmissions, both of which require air interface security, and from the reader to the enterprise, all middleware applications and user(s). The regulatory oversight of the communications chain is illustrated in Fig. 8.1 [5]. Security architectures can be constructed to address the numerous threats and countermeasures so as to mitigate the risk of compromising the data. Depending on the application, the level of impact could be negligible to cata-strophic for an organization, or even a nation. The Association for the Automatic Identification and Mobility's (AIM) RFID Experts Group (REG) has issued a document (AIM REG 314) on *RFID – Guidelines on Data Access Security* [5] that outlines the impacts, types of threats, and potential countermeasures to address (items 1 and 2 in Fig. 8.1) at the tag and the air interface for tag-to-tag or tag-to-reader communications. As illustrated (Fig. 8.1), there are ISO standards that are applicable to different segments of the RFID system architecture.

Interference and integration with other systems

Extreme care needs to be taken when deploying both passive and active RFID systems in environments where other RF systems are transmitting or receiving. Installing some systems in warehouses with older wireless networks has resulted in the wireless network being rendered inoperable. Other accounts of system failures have been observed when multiple RFID vendors were demonstrating their units. The interface or power emission from one system would shut down others. Just as important as interoperability between RFID systems and other wireless transmission systems is the need for testing the impact on other electrical systems that could be impacted, such as medical devices. Potential disruption of the functioning of a medical device by an RFID system field is of serious concern to the wearer of the device, the manufacturer of the device, and the manufacturer of the RFID system. Therefore, potential interactions between medical devices and RFID systems must be determined so that appropriate steps can be taken to minimize undesired electromagnetic environmental effects.

It is important to note that all devices and/or systems that operate at the same or similar frequencies have the potential to interfere with each other. Effective site surveys are required prior to the implementation of RFID applications in order to understand the current environment in which the RFID technology will be installed. There is already a good deal of knowledge that has been gained by the collection of early adopters of RFID on dealing with interference and frequency issues, as described in Chs. 10 and 11.

When installing a new RFID system in an environment that already has an RF system in use operating within the same frequency range, the only way to guarantee that no interference will occur is either to shut one system off or to change the frequency of one of the systems. If both systems are operating at a particular frequency, or at very similar frequencies, the best that can be done is to minimize the interference. The initial stages of an RFID pilot should be devoted to acquiring a proof-of-concept demonstration that includes functioning legacy and RFID systems. In addition, frequency coordination tests should be conducted before fielding the RFID systems.

One of the biggest sources of interference for RFID systems comes from legacy 900-MHz wireless local area networks (WLANs) that are currently being used in many warehouses.[2] There are ways to work around this issue by setting these systems up so that they do not use the same RF channels at the same time. It is also important to note that many of these older WLAN installations are being replaced by 2.45-GHz systems.

The ability to take advantage of existing wireless network infrastructure makes combining active RFID and WLANs more attractive in some instances. While RFID technology is unrelated to that of WiFi standard 802.11 (although there is an RFID specification in the 2.4-GHz band that is also used by 802.11b/g.), the communication link between the RFID reader and the in-building network can be the 802.11 WiFi network.[3] However, the WiFi network has to be designed to provide sufficient capacity and throughput and to have sufficient RF signal coverage to support RFID readers. In the instance of RTLS systems, as is

[2] See http://www.zebra.com/id/zebra/na/en/index/rfid/faqs/frequencies_interference.html.
[3] See http://www.connect802.com/rfid_facts.htm.

identified in Ch. 6, the denser access points required to achieve an acceptable accuracy level of localization must be analyzed as against separate infrastructures.

RFID signals are furthermore susceptible to environmental conditions – thick walls, nearby metal surfaces, the presence of liquids, static electricity, and electromagnetic induction. There are ways to work around these environmental constraints, including adjusting the density (or number and position) of access points (readers), the signal strength, and the location itself [6]. The main factors that counter interference include the size of the antenna, the frequency used, the power output, and the composition of the tagged objects. Methods to control interference include placing a buffer between the tag and the source of interference, altering the location of the antenna, and placing readers as far away as possible from sources of interference.

As with all new wireless technologies, early adopters of RFID have encountered issues from many unexpected sources. For example, IBM tested RFID equipment in the back-room grocery sections of (seven) pilot Wal-Mart stores, in support of the retailer's RFID project [7]. During the deployment, IBM consultants encountered interference from hand-held devices such as walkie-talkies, forklifts, and other devices typically found in distribution facilities. Nearby cell-phone towers, which transmit at the high end of the frequency band, sometimes leaked unwanted radio waves to the RFID readers. Bug zappers in the back rooms of the test stores also caused interference.

In another example IBM discontinued an RFID demonstration at the 2006 Australian Tennis Open after the radio RFID signal blocked calls from a nearby Vodafone phone tower [8]. The incident confirmed long-running concerns that the readers used to scan RFID tags could disrupt towers in Vodafone's GSM mobile phone network because of overlaps in the frequencies used by the two systems.

Retailers have faced challenges with RFID readability, interference, and consistency of tag performance [9]. Because many RFID readers are situated close to consumers, retailers must use low power, which restricts the ability to read over distances. The issue of whether the reader can read large volumes of information fast or accurately enough is a complex equation that is addressed in the technology chapters of this book. For example, Tesco cited problems with RF standards and the high number of readers in the warehouse as affecting performance, slow read rates, and intermittent tag quality. Del Monte tested RFID under warehouse conditions for 18 months, subjecting tags to interference from mobile phones and other electronic devices. The tags' ultra-high frequencies were impacted by interference from metals and liquid.

Radio frequency identification systems as used by the DoD to track and locate supplies have the further challenge of managing interference with – and potentially degradation of – critical radar systems [10]. The military services operate radar systems in the frequency band 420–450 MHz, so the potential exists for interference from active tags. One official we spoke to indicated that active battery-powered tags and readers that operate at 433 MHz caused degradation in radar performance. By contrast, passive tags operating in the frequency range

868–956 MHz did not cause similar degradation. These active RFID tags have a nominal range of 300 feet. To prevent interference from new, higher-powered tags and readers that operate with a 60-second operating cycle, the FCC has barred the use of 433-MHz active tag readers within 25 miles of key military radar systems used for missile tracking.

8.5   Testing for RFID performance and interference

Use cases for RFID testing

Although many professionals in the security services are convinced that RFID is a promising technology to pursue, some experts consider that there are still many variables in RFID implementation that require thorough testing and forethought before users deploy RFID systems [11]. RFID tag lifespan, readability rates, and interference from other sources are just a few aspects of RFID that require implementation-specific testing.

RFID tags will perform differently depending on the composition and contents of the container [11]. Multi-directional testing must be performed on the actual container to determine the right RFID smart label and best location for placement on the container to achieve optimal read performance.

Extensive testing is the key to successful deployments, as discussed by Kurt Menges in his article [12]. Interference can be isolated and measured much more conveniently in test laboratories than under actual conditions as a first step. Because not all of the variables that will affect performance can be anticipated in a laboratory setting, extensive on-site testing is also recommended.

Tag testing and validation include tag selection, tag placement, and product testing in the laboratory and subsequently in real distribution center environments. Tag antennas come in various sizes, shapes, and material compositions. The analytical objective is to find the tag that performs best with the product and packaging at the lowest possible price. The optimal placement of tags on product cartons or cases is affected by the carton's contents, by how they are packaged, and by the packing materials. Optimal tag placement must take into account other labeling requirements (i.e. not obscuring preprinted text and graphics).

Product testing for speed and distance readability (on conveyor belts and through dock door portals) is best accomplished in a real distribution center environment. Testing in a warehouse also enables the assessment of operational changes that may be required in order to accommodate the new RFID system. By contrast tag selection and placement testing can be performed in a controlled laboratory setting. However, prior to actually utilizing the RFID system, these tag variables will need to be retested as tagged products pass along the actual conveyor belt and through dock door read zones in the warehouse.

Before designing and installing portal and conveyor infrastructure, a site survey tests the warehouse RF environment itself for sources of interference. These sources of interference can be caused by various items of electrical equipment and wireless devices in use throughout the warehouse. To remove commonly found sources of interference, companies are finding that they need to upgrade their WLANs and barcode systems (that operate in the UHF spectrum) to updated technology that operates at non-interfering frequencies.

Testing should be performed during the busiest time at the location in order to get a true picture of the RF noise that is likely to occur. Each target read zone should be tested individually, to obtain an understanding of the interfering signal strength in each area. Finally, it is recommended that RF site survey testing lasts for a period of 24–48 hours (rather than taking a snapshot at a particular point in time). Once sources of interference have been identified, they can be either removed or worked around. Sometimes this can be as simple as moving the interfering device away from the read zone or shielding it.

Medical equipment testing

A considerable amount of testing of medical devices with respect to interference from commonly encountered electromagnetic systems, such as electronic article surveillance (EAS), has been conducted over the past 10–15 years; however, limited interference testing for RFID systems has occurred to date. Just recently the FDA, as presented at the Fifth RFID Academic Convocation in Orlando [13], and a few medical device manufacturers have performed some initial tests. Since HF RFID systems operate in the same frequency range as, or have similar modulation characteristics to, EAS systems in that frequency range, it is reasonable to assume that test results with those EAS systems may apply to HF RFID systems.

The maximum allowable transmitter output power for emitters, including RFID systems, is set by FCC and/or human exposure requirements. Since exposure standards regarding human safety are based upon average power and short-term exposure, the allowed peak power of pulsed signals such as RFID may exceed the immunity level of medical devices.

The FDA recommends that manufacturers of medical devices susceptible to RF interference address the selection of appropriate RF wireless communication and shielding technologies in their design as part of the risk-management process. This is due to the limited amount of RF spectrum available and potential competition among wireless technologies (including RFID) for the same spectrum.

The FDA recommends that electromagnetic compatibility (EMC) be an integral part of design, testing, and performance assessment for RF wireless medical devices. Voluntary consensus standards such as the IEC 60601-1-2:2001 *Medical Electrical Equipment – Part 1-2: General Requirements for Safety – Collateral Standard: Electromagnetic Compatibility – Requirements and Tests* provide requirements on electromagnetic emissions and immunity for electrical medical equipment.

A brief summary of examples of potentially problematic situations with medical equipment in the presence of RF technology includes the following situations.

- Electromagnetic interference (EMI) effects on implantable medical devices, including such responses as missed output pulses, noise reversion, rate changes, and inappropriate delivery of therapy.
- EMI causes erroneous wireless programming of a medical device.
- Too many nearby in-band RF transmitters increase the latency of, or slow down or block, transmission of continuous, critical patient data, and life-threatening conditions are not communicated to caregivers in a timely manner.
- RF wireless transmissions cause EMI in other nearby medical devices.

The FDA recommends that manufacturers design and test medical equipment for EMC performance using appropriate methods in light of these issues.

8.6   References

[1] **Q.E.D. Systems,** "RFID: Active, Passive and Other – Clarification," in *AIM Insights* (AIM global, Warrendale, PA, 2006) (http://www.aimglobal.org/members/news/templates/aiminsights.asp?articleid=2000&zoneid=26).

[2] **Harmon, C.,** *Active and Passive RFID: Two Distinct, But Complementary, Technologies for Real-Time Supply Chain Visibility* (http://www.autoid.org/2002_Documents/sc31_wg4/docs_501-520/520_18000-7_WhitePaper.pdf). This white paper was abridged by Q.E.D. Systems from two white papers created by Savi Technologies: *Active and Passive RFID* and *Selecting the Right Active Frequency.*

[3] **Bennett, G.,** "RFID and Smart Containers," *First RFID Academic Convocation,* Cambridge, MA, 2006.

[4] **Seshargiri, K. V., Nikitin, P. V.,** and **Lam, S. F.,** "Antenna Design for UHF RFID Tags: A Review and a Practical Application," *IEEE Transactions on Antennas and Propagation,* 53(12):3870–3876 (2005).

[5] **AIM RFID Experts Group,** *RFID – Guidelines on Data Access Security* (AIM, 2007) (http://www.aimglobal.org).

[6] **Carrasco, L.,** *The Convergence of Wi-Fi and RFID* (2005) (http://www.ebizq.net/topics/rfid/features/6328.html?page=2&pp=1).

[7] **Thillairajah, V., Gosain, S.,** and **Clarke, D.** *Realizing the Promise of RFID (Part II of II)* (2005) (http://www.ebizq.net/topics/rfid/features/6165.html?&pp=1).

[8] **Woodhead, B.,** *IBM Pulls Plug on RFID Demo* (2006) (http://australianit.news.com.au/articles/0,7204,20592544%5E15306,00.html).

[9] **Friedlos, D.,** *Testing Times for RFID Tags* (2006) (http://www.itweek.co.uk/computing/analysis/2159742/testing-times-rfid-tags).

[10] **Brewin, B.,** and **Tiboni, F.,** *Tests Raise Questions about RFID: Radio Tags May Interfere with Military Radar* (2005) (http://www.fcw.com/article90476-08-29-05-Print&printLayout).

[11] **Steward, S.,** *Preparing for RFID Lift-off: One Company's Pilot Program Tests RFID's Variables* (2005) (http://www.devicelink.com/pmpn/archive/05/12/011.html).

[12] **Menges, K.**, *Real World Experiences in Selection, Validation, and Deployment of RFID* (2006) (http://www.ebizitpa.org/RFIDconference/Articles/Real_World_Experiences_in_Selection.pdf).

[13] **Seidman, S.**, *et al.*, "Electromagnetic Compatibility of Pacemakers and Implantable Cardiac Defibrillators Exposed to RFID Readers," *Fifth RFID Academic Convocation*, Orlando (2007) (http://autoid.mit.edu/ConvocationFiles/Seidman-%20EMC%20of%20pacemakers.ppt).

9 RFID in the retail supply chain: issues and opportunities

Bill C. Hardgrave and Robert Miller

9.1 Introduction

In June 2003, Wal-Mart asked its top 100 suppliers to begin using RFID tags on pallets and cases shipped to the Dallas, Texas region by January 2005. This announcement, coupled with initiatives by companies such as Metro, Tesco, and Albertson's, propelled the RFID industry forward. In the period since that announcement, companies have worked feverishly to understand and utilize RFID in their supply chain in order to meet retailers' requirements and, more importantly, create business value for themselves. Initial RFID implementations predominately affected only a small portion of the supply chain (from retailer distribution center to store backroom) and focused on tagging pallets and cases. Given this limited supply chain exposure, determining the payback and, ultimately, creating business value have proven challenging. Early research is, however, providing evidence that, even under the aforementioned restrictive conditions, RFID is yielding a positive return: (1) Gillette found that RFID makes a difference in the tracking and managing of promotions [1]; (2) MIT's work on electronic proof-of-delivery (ePOD) at the retail distribution center revealed a valuable use of RFID [2]; and (3) results from a Wal-Mart-supported University of Arkansas study suggested that RFID can help reduce out-of-stocks [3]. Given the limited scope – i.e. retailer distribution center and store; pallets and cases only – it is very encouraging that solid business cases have already been found. In reality, these preliminary uses of RFID in the supply chain are merely the tip of the proverbial iceberg. The biggest gains are yet to be realized. However, with the bigger gains come bigger challenges. In this chapter, we discuss the challenges and opportunities facing the RFID-enabled supply chain as it expands both geographically and to encompass greater portions of the supply chain, including item-level tagging.

9.2 From partial to full supply chain coverage

As mentioned, RFID's use within the supply chain has been limited to primarily the retailer DC and retail store back room. Although business value has been

RFID Technology and Applications, eds. Stephen B. Miles, Sanjay E. Sharma, and John R. Williams. Published by Cambridge University Press. © Cambridge University Press 2008.

found, using RFID in an abbreviated supply chain (often called "slap-and-ship") limits the potential benefits and, in fact, puts most of the pressure to show a return on investment on the retailer. To reach its full potential, RFID must be used throughout the supply chain – from the farm/factory to the consumer. What is the value of tagging and tracking products from the beginning of the supply chain (wherever that might be) to the retail store? One answer to this question is that RFID could be used to support a system of provenance that would allow parties along the supply chain to easily identify all past points in the chain. As an example, RFID would make it much easier to trace the origin of contaminated food lots. As another example, the pharmaceutical industry estimates that between 2% and 7% of all drugs sold globally are counterfeit [4]. In response to growing concerns, the FDA Counterfeit Drug Task Force has instituted a requirement that pharmaceutical manufacturers maintain records on the history of a drug's buyers and sellers throughout the supply chain (from manufacturer to consumer) [5].

This ability to "see" products along the supply chain provides additional insight into two key metrics (among many): shrinkage (theft, damage, etc.) and lead times. Although most companies have an idea of the amount of shrinkage, determining the exact amount and exactly where it occurs has proven elusive. With the visibility provided by RFID, companies can have a much better idea as to how much shrinkage there is and where it occurs along the supply chain. Currently, lead time estimates (how long it takes to move product between various points in the supply chain) are imprecise. Because lead times are used to determine reorder points and safety stock, for example, accuracy is important. With RFID, companies can do a much better job of determining lead times due to the precision of tracking at the case level and establishing exact times of arrival and departure at key points in the supply chain.[1]

This level of visibility into product movement along the supply chain is based on the ability to actually read the RFID tags at key points. RFID is not perfect – passive UHF systems (as used currently in the retail supply chain) have problems with water and metal (with absorption and reflection of RF, respectively, as indicated in other chapters in this book). As one of the four performance test centers (PTCs) originally accredited by EPCglobal (and the only academic lab), we have had the opportunity to perform numerous tests of consumer packaged goods products for readability. As a PTC, we help companies find the best tag type and tag location for their product. Do tag type and tag location matter? In our experience, they certainly do. Figure 9.1 shows an example (from an actual test engagement) of how much the read rates can vary depending on tag type/tag placement.

In this example, seven different tag type/tag location combinations were tried. As shown, read rates ranged from 10% to 100%. Thus, tag type/tag location does matter. During the first full year of testing after converting to Generation 2 equipment, we were able to get 100% read rates on every product we tested.

---

[1] For more information about the determination of lead times with RFID, see [6].

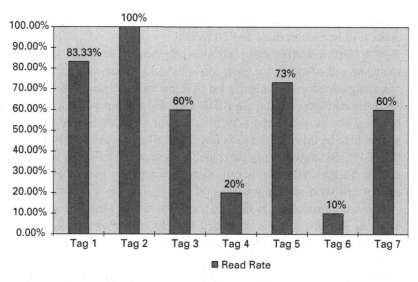

Fig. 9.1 The importance of tag type/placement for read rates.

Overall, companies must be cognizant of the need to perform proper tag type/tag location testing prior to deployment.

Expanding the use of RFID throughout the supply chain necessarily involves a geographic expansion, since manufacturers are spread across the globe. The global use of RFID is not without its challenges. First, different countries have different standards for RFID. This issue continues to improve, but must be resolved before RFID can become globally accepted. China is a major player in the global market from a manufacturing perspective and their adoption of RFID standards, rather than trying to establish their own, is important. Second, not all supply chains are the same. In this chapter, we have taken a very narrow view of what a supply chain is and have used the term generically. In reality, there are many different supply chains, each with its own benefits and challenges. For example, shipping a case of diapers through the supply chain is very different from shipping a case of frozen turkeys. Also, supply chains tend to vary by global region. For example, in the USA, retailers are geographically dispersed, requiring less frequent, but larger shipments; whereas in Europe, retailers are not as geographically dispersed, which allows the use of more frequent, but smaller, shipments. Thus the business case for RFID may vary by supply chain and by global region.

9.3   Store execution

Improving the global supply chain helps get the product to the store efficiently and effectively. But what then? The "last 50 feet" – from back room to shelf – often causes the most problems in the retail supply chain. RFID is already starting to make a difference in these last 50 feet.

Wal-Mart and other major retailers rely heavily on visual inspection to determine an out-of-stock (or near out-of-stock) situation. These items are then manually added to a picklist (a list of items to be retrieved from the back room). By using data generated from RFID showing when product arrived and whether or not it has moved to the sales floor, retailers can determine whether product is available to be taken to the sales floor (and whether it should be taken to the sales floor) [3]. By automating the picklist data with RFID reads, Wal-Mart has achieved a 30% reduction in out-of-stocks simply by using RFID-tagged cases to improve its shelf-stocking process [7].

The same process as is used to create better picklists for shelf stocking (and reducing out-of-stocks) is also used to help reduce unnecessary manual orders. When a store employee, mistakenly thinking that a product is out of stock on the shelf and in the back room, manually orders a product, an incorrect signal of demand (and supply) is sent upwards through the supply chain. This signal, being based on inaccurate information, causes unnecessary fluctuations (often referred to as the bullwhip effect). With RFID, Wal-Mart was able to reduce unnecessary manual orders by 10% [8].

Another area where store execution has been improved with RFID is in promotions. A "promoted" product is one that is being emphasized through such means as advertising and coupons. According to research from the National Association for Retail Merchandising Services, about 56% of promotions are not properly executed nationwide [9]. RFID can provide visibility into the movement of promotional items or promotional displays into the store and onto the sales floor – visibility that is key to understanding and determining proper store execution of promotions. As evidence of the usefulness of RFID in promotions, Gillette found a 19% increase in promotional sales due to the use of RFID [1] [9].

To date, the advantages of RFID in store execution rely upon the tagging of cases and the visibility provided by readers at the back of the store (e.g. inbound doors, storage racks, transition doors from the back room to the sales floor, and box crushers). However, since Wal-Mart's announcement in June 2003 of a limited rollout of RFID at the pallet and case level, perhaps the question asked most often is "When will Wal-Mart (and other retailers) start using RFID at the item level?" Although it will probably be several more years before the use of RFID on supermarket items, such as a can of soup, becomes ubiquitous, the field has advanced rapidly in the past few years. In fact, extended pilots involving item-level tagging are well under way [10]. For example, Levi Strauss, Japanese retailer Mitsukoshi, and UK retailer Marks & Spencer are tracking apparel and footwear; Tiffany & Co., GN Diamond jewelry wholesaler, and Swiss watchmaker/jeweler de Grisogono are tracking watches and jewelry; and Yodobashi Camera is tracking digital cameras [10]. Within the upscale retail goods sector, organizations see RFID item-level tagging as one potential way to combat counterfeiting and improve on-shelf availability [10]. With item-level tagging, the utopian dream of the "contactless check-out," whereby a customer can simply carry (or push) their purchases out the door (legally!), without standing in line, is within reach.

For almost any type of retailer, item-level tagging provides perhaps the best opportunity for payback, given the limited visibility most retailers have of merchandise on the store shelves. Let's examine one simple area of opportunity: inventory. Although most retailers calculate a perpetual inventory based on point-of-sale data, these counts are inherently wrong – about 65% of these counts are inaccurate [11]. Consider an all-too-familiar scenario: a customer purchases six individual cans of green beans – three with salt, three without salt. The cashier scans one can, then, to save time, simply enters "X 6" to record the scan six times. Unfortunately, in an effort to save time, she has caused the perpetual inventory count to be wrong on two products (one will show three more cans than it actually has, one will show three cans fewer). RFID would attenuate this type of error. To correct inventory counts currently, retailers must physically count their inventory. Conducting physical inventory counts is time-consuming and expensive. Taking a physical inventory count with RFID would be much easier – imagine the shelves (equipped with RFID readers) taking an inventory for you every night! By knowing precisely what is on the shelf (via a correct perpetual inventory count) and what is in the back room (via case-level tagging), retailers could reduce the incidence of out-of-stocks, thus increasing overall sales and customer satisfaction.

The ubiquitous use of item-level RFID faces several significant hurdles. For item level, cost concerns become increasingly more important (due to the large number of things that must be tagged) and more difficult to address. For many consumer packaged goods, profit margins are very thin. It currently costs only about 0.1¢ to print a barcode on a product (e.g. a bottle of ketchup). Although RFID tag prices have dropped dramatically in the past few years, they are still much too expensive (about 10 cents) for typical supermarket products. To lower the costs, alternatives to the existing tag type (i.e. a silicon chip with a metallic strap of some type) must be considered. For example, several companies, such as Philips Electronics, are working on a printable RFID tag using electromagnetic ink. This type of technology can help facilitate item-level tagging for many low-profit-margin items. It can also help avoid other issues of using silicon and metal strapping on packages of food products. Currently, the metal in RFID tags prevents their use on many products due to the sensitivity of metal-detecting machines in the packing process.

Another technology-related challenge involves standards. The current retail supply chain uses passive UHF. Given the problems with UHF (i.e. those caused by water and metal), how can we tag individual items such as a can of chicken noodle soup or a bottle of water? The answer, some believe, is to use HF tags. However, compared with UHF, HF requires different systems architecture – different readers, antennas, etc. For retailers, this would mean one architecture for pallet and case (based on UHF) and another for items (based on HF). Dual architectures, obviously, are not ideal. Impinj recently introduced near-field UHF, which uses a standard UHF reader, but different antennas, to capitalize on the existing UHF infrastructures [12]. Another technology, dubbed "RuBee," is also making a play in the item-level arena [13]. Overall, the issue of competing standards must be addressed. Perhaps the best solution would be a technology one – a single architecture that is capable of reading many different types of tags.

9.4   Data analytics

The successful diffusion of RFID throughout the entire supply chain (including the store) depends on the data generated. Simply deploying RFID provides little, if any, value for an organization. Rather, the true value of RFID lies in the data it produces. Even at this nascent stage of deployment, RFID is already providing unprecedented visibility into the movement of products. Wal-Mart suppliers, for example, have near real-time visibility as read data is provided via Wal-Mart's RetailLink system (an extranet providing point-of-sale and now RFID data). This has enabled suppliers to see when their products reach the distribution center and the stores. Suppliers can even see the movement of their products among various read points within the stores. Key metrics, such as lead times and shrinkage, can begin to be determined with this data. Wal-Mart is setting the example on how to share the data with their supply chain partners. For RFID to be useful, all supply chain partners must be willing to participate by sharing their RFID data. As the use of RFID within the supply chain expands, additional data (and associated visibility) will be generated. More research is needed in order to understand and interpret the data that is generated from an RFID-enabled supply chain (including at item level). Understanding the data is the key to producing business value.

Given the amount of merchandise flowing through major retail supply chains (such as Wal-Mart), the RFID-enabled supply chain will generate a large amount of data. However, it is not the size of the data files that will create problems; it is the number of records (or data points) generated [14]. As an example, a company using RFID within an abbreviated supply chain that moves about one million cases daily would generate several million records a day [14]. While hardly overwhelming for a modern information system, this number of records would still be substantial. The challenge for companies using RFID is that of how to mine valuable information from the large quantity of records produced from an RFID-enabled supply chain. Also, keep in mind that the amount of data will increase exponentially with item-level tagging.

The RFID data currently generated is not perfect. Rather, there are often numerous duplicate reads, per product, per read point (e.g. case 123 could be sitting on a pallet at a dock door for several minutes and generate thousands of duplicate reads). Who is responsible for filtering this data: the retailer or the supplier? Most systems have rudimentary filters built in at the source (e.g. the retailer), but it is still possible for a massive number of duplicate reads to get through. For example, one anonymous supplier reported receiving more than

19,000 reads from one case at one read point from Wal-Mart [10]. Thus, suppliers must be diligent in removing unwanted, duplicate reads from their data before processing. In addition, the data must also be cleaned of "inadvertent reads." These are reads that should not have occurred, but were captured because the product was taken near a read point inadvertently. As an example, a non-empty case that is being moved through the back room of a store may inadvertently be read by the reader at the box crusher. Suppliers must be aware of such anomalies in the data so that the data can be cleaned accordingly. To fully understand RFID data, supply chain partners must be willing to share business processes – otherwise, it would be difficult (if not impossible) to understand important patterns in the data.

Overall, RFID provides the opportunity for near real-time, unprecedented, visibility into the supply chain. There is much to discover from the data and, ultimately, the data holds the key to business value. Although RFID technology is advancing rapidly, the use of the ensuing data is not keeping pace. Companies must be able to absorb and utilize the new data that will be provided. Key metrics, such as lead times, determined from RFID data will begin to pave the way for additional, not-yet-imagined, metrics in the future.

9.5   Conclusion

In this chapter, we have examined key areas of opportunity and challenges with the expanded use of RFID in the supply chain. To date, efforts have been focused on an abbreviated supply chain (retailer's distribution center through store back room) at the pallet and case level. While much has been learned from this limited rollout, the best is yet to come. The use of RFID at the item level offers the advantage of total visibility of one's products inside the store – which has never before been possible with barcodes. The global expansion of RFID throughout the supply chain promises to improve product traceability, while driving out inefficiencies. These RFID initiatives will provide an unprecedented ability to improve the supply chain. However, the success of RFID relies upon the proper use of its data. There is still much work to be done in understanding and interpreting this significant business resource.

9.6   References

[1] **Murphy, C.**, "Real-world RFID: Wal-Mart, Gillette, and Others Share What They're Learning," *InformationWeek* (May 25, 2005) (http://informationweek.com/story/show Article.jhtml?articleID=163700955&_loopback=1).
[2] **EPCglobal**, *Electronic Proof of Delivery* (2006) (http://www.epcglobalinc.org/news/ EPODVignetteApprovedV2.pdf).

[3] **Hardgrave, B., Waller, M.,** and **Miller, R.,** *Does RFID Reduce Out of Stocks? A Preliminary Analysis* (Information Technology Research Institute, Sam M. Walton College of Business, University of Arkansas, 2005) (http://itrc.uark.edu/research/display.asp?article=ITRI-WP058-1105).

[4] **Whiting, R.,** "Drug Makers 'Jumpstart' RFID Tagging of Bottles," *InformationWeek* (July 26, 2004).

[5] **Feemster, R.,** "FDA Raises the Stakes," *Pharmaceutical Executive*, 26(7):54 (2006).

[6] **Delen, D., Hardgrave, B.,** and **Sharda, R.,** "RFID for Better Supply-chain Management Through Enhanced Information Visibility," *Production and Operations Management*, forthcoming (2007).

[7] **Hardgrave, B., Waller, M.,** and **Miller, R.,** *RFID Impact on Out of Stocks: A Sales Velocity Analysis* (Information Technology Research Institute, Sam M. Walton College of Business, University of Arkansas, 2006) (http://itri.uark.edu/research/display.asp?article=ITRI-WP068-0606).

[8] **Sullivan, L.,** "Wal-Mart RFID Trial Shows 16% Reduction in Product Stock-outs," *InformationWeek* (October 14, 2005).

[9] **Shermach, K.,** "Proving the ROI for RFID," *TechNewsWorld* (July 12, 2006) (http://www.technewsworld.com/story/51662.html).

[10] **Roberti, M.,** "Using RFID at Item Level," *Chain Store Age*, 82(7):56 (2006).

[11] **Raman, A., DeHoratius, N.,** and **Ton, Z.,** "Execution: The Missing Link in Retail Operations," *California Management Review*, 43(3):136–152 (2001).

[12] **O'Connor, M.,** "Wal-Mart Seeks UHF for Item-level, *RFID Journal* (March 30, 2006) (http://www.rfidjournal.com/article/articleprint/2228/-1/1).

[13] **Schuman, E.,** "RuBee Offers an Alternative to RFID," *eWeek.com* (June 9, 2006) (http://www.eweek.com/article2/0,1895,1974931,00.asp).

[14] **Hardgrave, B.,** and **Miller, R.,** "The Myths and Realities of RFID," *International Journal of Global Logistics & Supply Chain Management*, 1(1):1–16 (2006).

[15] **Roberti, M.,** "*The Mood of the EPCglobal Community*" (2005) (http://www. rfidjournal.com/article/articleview/1875/1/2/).

10 Reducing barriers to ID system adoption in the aerospace industry: the aerospace ID technologies program

Duncan McFarlane, Alan Thorne, Mark Harrison, and Victor Prodonoff Jr.

10.1 Introduction

The five years from 2000 saw enormous developments in the way in which technologies such as RFID could be deployed in the consumer goods supply chain as illustrated in the preceding chapter (Ch. 9). While many of these developments were generic, it became increasingly clear that other sectors would need to make substantial adjustments were they to capitalize on the significant cost reductions and standards developments that had occurred. It was for this reason that the Auto ID Labs set up the Aerospace ID Programme. The aim of the programme was

To remove barriers to widescale automated ID deployment in the aerospace sector through timely and effective R&D.

The barriers to be examined ranged from issues of technical feasibility, via economic viability hurdles, to questions of operational *viability* – that is, whether solutions could survive a harsh range of operating conditions. These hurdles to be addressed (see Fig. 10.1) served as a sanity check for setting the research directions which are reported in Section 10.5.

This chapter tells the story of the Aerospace ID Programme, its formation, its operations, and the results.

10.2 Background

As mentioned above, the background to the Aero ID Programme was the major development in the use of RFID in the consumer goods industry, led by the Auto ID Center and exemplified by the major initiative from Wal-Mart in 2004.

RFID Technology and Applications, eds. Stephen B. Miles, Sanjay E. Sharma, and John R. Williams. Published by Cambridge University Press. © Cambridge University Press 2008.

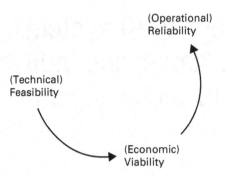

Fig. 10.1   Barriers to widescale ID systems deployment.

Table 10.1. Differences between consumer goods and aerospace RFID environments

	Consumer Goods	Aerospace
Product lifetime	Consumer goods have lives measured in months	Typical aircraft component may have a life of 20–50 years
Product characteristics	Low value and low complexity	High value and high complexity
Key application areas	Supply chain logistics applications are critical	After-sales, product service, repair and spares management (in addition to logistics)
Environmental conditions	Consumer goods supply chains are relatively well controlled	Aerospace components are often subjected to fiercely varying environmental conditions

However, in parallel with these developments, work to examine the possibilities of widespread deployment of technologies such as RFID in commercial aerospace had begun and in particular Boeing and Airbus ran a series of global forums to bring together their common supply base to consider the challenges involved in deploying RFID on aircraft. Through discussions such as these and early trials a number of key differences with the consumer goods case for RFID began to emerge. In Table 10.1 some of the product and environmental differences between the sectors are highlighted. A typical aerospace component has up to 50 years of life, may be constructed from hundreds or thousands of subcomponents, can have a value of up to $1 million, and is subjected to stringent monitoring throughout its life.

Such requirements can fundamentally alter the RFID system's functionality and, as shown in Fig. 10.2, this is illustrated in some early assessments of the likely tag complexity for different applications.

In 2004 when the planning for the Aerospace ID Programme began, the environment was not especially conducive to widespread deployment of RFID in the aerospace sector. The few reported application trials were typically single bespoke applications, where vendors were working with a single company or conducting

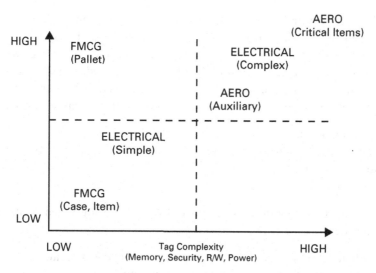

Fig. 10.2   Assessment of tag complexity for different applications.

private cluster trials. Efforts in standardization and R&D at the time were wholly focused on low-cost passive solutions, low-memory tags, and consumer goods supply chain challenges. Hence an end-user-driven, cross-supply-chain, deployment-oriented program was formulated with the intention of identifying standard types of solutions for the industry. In particular it was intended that by 2007 the program would have assisted the aerospace industry in achieving the specification of clearly specified, cost-effective identification technologies, including

- a logical mix of different ID technologies;
- low- and high-class RFID;
- multi-company pilots, including public demonstrators with openly accessible results;
- understood solutions for integrating tags and sensors, and managing multi-dimensioned product data, with integrated network management and support (interfaces);

in addition to the existence of

- vendors focused on the aerospace industry and its specific needs; and
- an ongoing network of academic/industrial R&D.

This vision required the involvement of a broad range of participants from both industrial and academic domains.

10.3   The Aero ID consortium

The consortium supporting the proposed research program was critical for the success of its mission. Consequently, significant effort was devoted to identifying suitable sponsors and determining their specific requirements. Three requirements workshops were held in 2004/5 – in Cambridge, in Berlin, and at the Paris Air Show – and in addition numerous company visits were made. The program was very fortunate to receive seedcorn funding from the Cambridge MIT Institute, which supported the preparatory activities. Discussions were held with over 100 companies and it was agreed that four key categories of partners would be required:

- industrial end users
- solutions providers
- industry associations
- universities

The main benefits offered to industrial partners being asked to invest in the program were

- the ability to influence the formation of the research program
- immediate and deep access to key research findings via
  - tools to support deployment
  - guidelines on complex requirements
  - demonstrations to clarify application issues

Table 10.2. Aero ID partners

END USERS
Boeing
Airbus
Embraer
Aviall
BAE Systems
Messier-Dowty

INDUSTRY BODIES
SITA
IATA
ATA
GS1/EPCglobal

SOLUTION PROVIDERS
BT Auto ID Services
AERO ID Ltd.
T-Systems
Afilias
Intelleflex
Systems Planning Corporation
VI Agents

UNIVERSITIES
Cambridge, UK
Keio, Japan
ICU, South Korea
Korean Aviation University, South Korea
Adelaide, Australia
EPFL, Switzerland

- early access to industry white papers
- networking opportunities
- the ability to participate directly in the R&D program and linked pilot trials
- the ability to work with key bodies to influence critical standards for ID and data management

By the time of its launch in September 2005, the consortium had 10 members, and it grew to a final size of 20 (in February 2007). The consortium membership is detailed in Table 10.2.

In addition to sponsorship fees, each partner agreed to provide active input into each of the research themes and to support the dissemination and adoption activities initiated by the program. These amounted to a significant investment in time and were critical in defining and executing the program. We discuss these issues next.

10.4   Defining a research program

Program structure

The Aero ID research program was designed to produce results in 6 months and to be complete in 18 months, with results being immediately taken up by the industrial partners. This was challenging – especially given the typical university lead times for starting and undertaking research – and required the following.

- An unconventional approach to R&D: the program needed a balanced portfolio of long-term views on core issues while also being highly reactive to short-term actions. As the industry evolved quickly, it was necessary to be able to adapt the program to changing requirements and priorities, as well as adapting to significant resource challenges.
- Partner involvement: a key intent was that of deeply integrating industrial collaborators in research and encouraging collaboration between partners on key topics.
- Planning the end: a succession plan was developed for each of the themes in order to maximize take up and the researchers worked closely with standards/industrial bodies seeking to adopt the results.

A conceptual structure of the program is given in Fig. 10.3, which shows the program arranged around a number of central themes, and the resources for the program being sourced from many different organizations, including sponsors.

In order to ensure rapid dissemination and take up of the results, the outputs were designed to be readily usable in the form of

- tools to support deployment
- analysis of pilot results

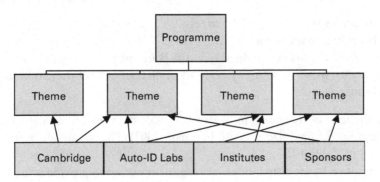

Fig. 10.3   Conceptual structure of the program.

- guidelines on complex requirements
- demonstrations to clarify application issues
- software prototypes
- industry white papers

Program content

As discussed earlier, the final decisions on the content of the program were drawn from a series of industrial consultations and reflected those key barriers to ID system adoption that could reasonably be addressed by a research program. Other issues such as standards formation and simplification and the need to establish sound business cases were also seen as critical but outside the scope of the program.

The main issues identified for investigation were as follows.

- *Lifecycle ID and data management strategies.* Given that aerospace components have an extremely long lifecycle, and that data entries are held on a variety of databases and with differing locator IDs, there was a need to develop a long-term model for secure lifecycle data management.
- *Tag and database synchronization.* The industry's interest in read/write tags required a strategy for synchronizing data maintained in duplicate on networked databases and on the tag.
- *Resilient tags and appropriate ID solutions.* The long lifecycle, coupled with exposure to many different environments, requires a robust, resilient tag, a long-term model for data management and security, and potentially the need for multiple ID solutions for a single component.
- *ID-sensor fusion.* Comprehensive management of an aerospace component requires that the selected ID solutions be combined with many other sensing devices in an effective manner. In particular, vehicle health management is a key application area for auto ID technologies in the aerospace sector.
- *Auto ID-enhanced track and trace.* The accurate and timely tracking of a component is critical in commercial and military aerospace applications.

Furthermore, the complexity of many components means that they contain many subcomponents and subassemblies, each with its own ID and replacement, repair, and upgrade processes. The main focus is on development and trialling of a model for specifying track-and-trace requirements appropriately and cost-effectively.

10.5   Research developments

Overview of developments

A principle established from the onset of the Aerospace ID Programme is that research be driven by the end users. From initial consultations the five themes presented in Section 10.4 were established and scoped. It was noted that significant interconnections between the themes existed and a mapping analysis ensured that relationships were understood and that there were no significant gaps or overlaps (as shown for example in Fig. 10.4).

Further to raising these issues, sponsors were also asked to provide internal contacts, with whom case studies could be conducted. A great number of such on-site studies allowed aerospace ID researchers to gain first-hand understanding of the particularities around aerospace processes and industrial operations.

The next step was to combine the insights from case-study findings with knowledge previously gained from work in other industrial sectors (notably in retail), as well as with fresh research being conducted around specific aerospace issues. The results have been divulged in 12 reports that are helping shape the aerospace industry landscape for the utilization standardized automatic identification.

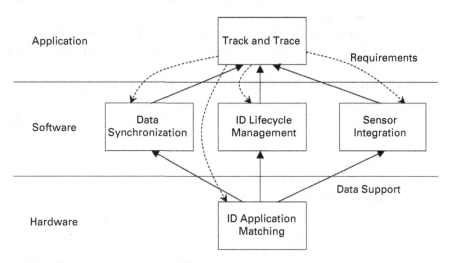

Fig. 10.4   Relationships between aerospace ID research themes.

The final logical step has been to validate research results in real-world trials back within the sponsors' operations. The current status of these trials is described in detail in Section 10.6. The following sections present in more detail the work that has been carried out on each of the research themes, explaining the most important results achieved along the path.

Theme 1 – ID application matching

During the formation stage of the ID technologies program, a considerable amount of time and effort was spent talking to different organizations across the aerospace sector. These organizations included manufacturers of airframes, engines, and avionics, as well as aircraft operators and maintenance providers. The diversity of ID technologies being used in different applications soon became apparent and a research strand was required to help address the issues involved when choosing appropriate ID technologies for different aerospace applications across different stages of a product's life.

The research strand highlighted the benefits and obstacles that different ID technologies bring to applications in the aerospace sector. It investigated the criteria and measures that are important in choosing an ID technology and then developed a software tool to help engineers quickly identify appropriate ID technologies for particular application requirements.

The decision process involved in choosing an appropriate ID solution for a particular application is complex and there are potentially many different solutions that would provide a working system. The designer is required to consider – amongst others – the questions below.

(a) What identity standards are currently in use?
(b) What are the legal/legislative/regulatory issues involved with using a particular identification technology?
(c) What are the technical issues involved with using a particular identification technology?
(d) What is the other data required to support the decision in the business process?
(e) Which physical processes are involved at the point where the identity is to be obtained?
(f) Are there environmental issues that the identification technology would have to handle?
(g) What are the costs/benefits that the identity information brings to the business process?

When considering even the basic questions above, it can be seen that the ID technology solutions workspace is a complex array of interlinked business processes, technology capabilities, and legislative requirements. To help start to simplify this workspace, they can be categorized using three basic performance measures – namely timeliness, accuracy, and completeness of the information available.

Using these measures it is possible to make some comparisons between different ID technologies using the three quality dimensions associated with product data.

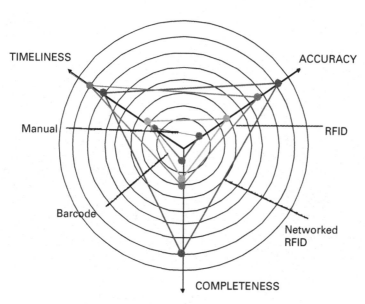

Fig. 10.5   Dimensions describing ID data quality.

The diagram in Fig. 10.5 shows the characteristics of manual data entry systems, barcode systems, RFID systems, and networked RFID systems.

When choosing an ID technology for a specific process, the quality measures described above provide guidelines about the nature of the data that the ID solution will provide. However, the physical characteristics of environments are important for the operation of ID technologies in the aerospace sector. This includes areas such as ID read range, RF absorption in material, and cost. Numerous further issues need to be considered in order to develop a comprehensive assessment of the most appropriate technology.

The research developed a model for guiding a system designer through the key decisions involved in specifying an ID system. The main output of the work is an ID application software tool (see Fig. 10.6), which allows engineers to go through an iterative process, providing specifications about the process and its environmental factors. The tool then highlights potentially suitable ID technologies that meet these requirements. (The tool has the ability to capture results from live trials so that they can be examined within a process context, rather than relying exclusively on the traditional ID specification sheet.)

Sensor integration

This research activity examined the rationale and approach for the integration of RFID and sensor data within the aerospace sector, because applications in complex part tracking and component health management necessarily require that identity and product state information be collected synchronously. The effective

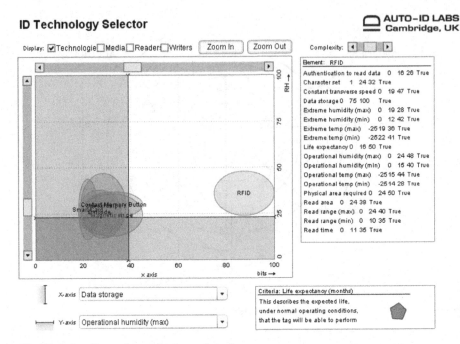

Fig. 10.6   The ID technology selector tool user interface.

combination of ID and sensory data in an aerospace context enables a wide range of queries to be performed on individual parts or platforms.

Table 10.3 gives simple and illustrative examples of queries that can be addressed if integration is done. (See Patkai *et al.*, 2006 for further details.) The need for a research program arose because of the limited attention this area had

Table 10.3. ID-sensor-enabled queries

	When the part has **RFID**	When the object has a **sensor**	When the object has both **RFID and sensor(s)**
Temperature sensor	*Where has the part been in the last month?*	*What is the temperature of the part?*	*Has the part been exposed to excessive heat in the last month?*
Vibration sensor	*Did a certain airline use this part last December?*	*Is the part vibrating due to a fault?*	*Has any specific part been vibrating **besides this part** due to a fault when it was owned by a certain airline last year?*
Humidity sensor	*Has this part been repaired more often than other parts of the same kind?*	*What is the humidity in the storage facility of a specific airline?*	*Has this part been repaired more often than other parts of the same kind because the humidity in the storage facility of a certain airline is usually high?*

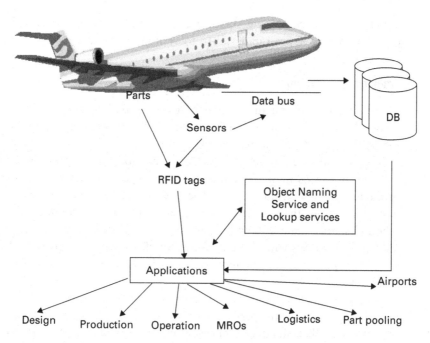

Fig. 10.7   ID–sensor applications.

received in the consumer goods domain, and two key areas for development were identified:

- ID and sensor data needs to be efficiently integrated into real-time data-capture systems;
- the potential of combining ID technologies with other sensory information (for example, ID and temperature, ID and location) had yet to be properly exploited.

In the aerospace domain, ID and sensor integration are critically important because of the long lifecycles of parts, the complex service processes and legislations in place, and the difficult operating environments that aircraft are exposed to (see Fig. 10.7).

In particular, it has been noted that the integration of RFID and sensors can be done in at least two distinct ways.

- Hardware integration. The sensor(s) is (are) connected physically to the RFID tag, and sensor data is read by the RFID reader.
- Logical integration. Sensor data is collected independently of the RFID tag, and the integration process involves the reading of the RFID tag and accessing another data source.

The selection of one of these routes is dependent on environment, cost, and the application. The research has also determined appropriate information architectures for combining ID and sensor data and case studies within partner organizations have been undertaken to verify these results.

Tracking and tracing

This theme focused on the process of tracking the current location of an item across the aerospace supply chain, as well as the process of keeping records and the tracing of a part's entire lifecycle history.

A series of case studies conducted at member companies revealed several business processes in which auto-ID technologies – and RFID in particular – could have a significant impact. The case studies set out to sample as much as possible from the spectrum of processes around a part's lifecycle, from the inbound supply chain linking first-tier suppliers to aircraft manufacturers, across industrial processes within the aircraft manufacturers themselves, and on to the after-sales processes of maintenance, repair, and overhaul (including reverse logistics).

Following this first-hand evaluation of aerospace processes, the theme delivered methods for assessing the performance of a tracking and tracing system and a way to estimate the benefits that stem from the deployment of an ID-based track-and-trace solution. The first part of this work was to develop a framework comprising a set of well-defined metrics that reflect the performance of a track-and-trace system. Some of the factors considered at this stage were timeliness and accuracy of information. The framework also included a standard way of applying these metrics in order to produce an objective and comparative assessment of the tracking and tracing solution. (See Fig. 10.8.)

Within this framework, a way to quantify the benefits that a company could get from a tracking solution was proposed, taking into account how improving the quality of tracking information could be used to improve business decisions and operations. This method may be used as part of an overall method for calculating the return on investment (ROI) from tracking and tracing solutions not only in aerospace, but also in other industries.

Lifecycle data management

The goal of using auto-ID technologies for identification of aircraft parts is to make it easier to gather more complete information about each individual part throughout its entire lifecycle. As discussed earlier, the lifecycle of an aircraft part is very different, for example, from the lifecycle of a consumer product in the retail supply chain. First of all, the parts are generally of much higher value than typical consumer products. Secondly, human safety depends upon checking that they are genuine parts, can be certified as being airworthy, and have been maintained correctly. Thirdly, whereas tracking of consumer goods often stops at the point of sale or transfer to the consumer, this is really only the beginning of the service lifecycle for an aircraft part, which may extend over

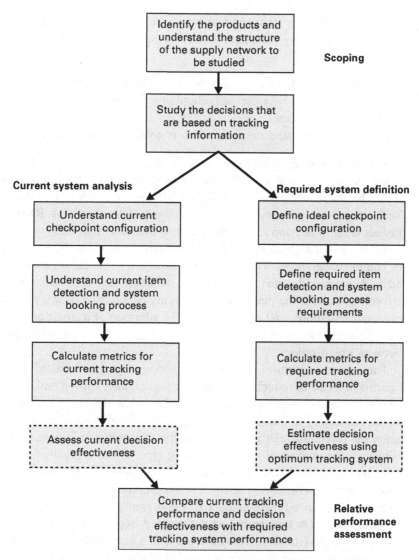

Fig. 10.8   A track and trace framework (from Kelepouris *et al.*, 2007, with permission) [13].

several decades and involve several changes of ownership, changes of custody, removal and installation from multiple aircraft, and repairs and upgrades at various times throughout its service life. No two parts are identical, because they each have an individual life history, which must be tracked and recorded separately for each part. To enable this, one essential requirement is a globally unique identifier, so that each part has a unique name that all players use to refer to that part. A further essential requirement is the ability to share and gather complete lifecycle information about the part from multiple providers of that information.

In 1992, the Air Transport Association (ATA) approved standards for aircraft-part-marking practices. The ATA Spec 2000 identifier consists of two components:

(i) a 5-character CAGE[1] code that unambiguously identifies a particular organization (usually the manufacturer of a new part – or a supplier of an in-service part); and

(ii) a 15-character serial number, which is unique within the CAGE code.

Today, the Spec 2000 identifier is widely used and can be found in part-marking name plates, expressed as alphanumeric human-readable markings, as well as in linear barcodes or two-dimensional data matrix barcodes. The Aerospace ID Technologies programme has worked closely with the ATA RFID on Parts project team to develop a proposal for how the same Spec 2000 unique identifier could be encoded on an RFID tag. Many aerospace organizations use a number of computer databases within their organization, some dating back several decades. A unique identifier is useful as a database key for retrieving additional information about the part, such as warranty details, and details of maintenance events, installations, and removals etc. The adoption of a single unique identifier and internal data integration is the first step to enable information sharing between organizations.

The communication of an electronic message in a standardized format brings benefits in terms of reducing the need for human intervention and errors that are caused by this, as well as resulting in swifter communication between organizations. The ability to improve a design process or manufacturing process depends upon the recording, alerting, and detection of systematic failures or weaknesses across a large number of parts of the same type, so this is greatly facilitated if the information is available in a standard machine-readable format for ease of comparison and correlation analysis – and especially if the lifecycle event data is communicated in a timely manner to the organizations that require such information. The ATA has developed a Common Support Data Dictionary, which codifies various data elements associated with a part, such that each data element can be encoded in an electronic message in a standard format.

Technologies such as high-capacity RFID tags and contact memory buttons offer the ability to store significant amounts of lifecycle data with the part, so that it can be read anywhere, without requiring a network lookup. However, this approach on its own does not provide the full benefits that are desired, since data stored on the tag or memory button can be read only when the part is physically present (usually by only one organization at any given time) – and then only if the tag or memory button has not already failed due to physical stresses (such as

---

[1] CAGE – Commercial And Government Entity – a company identifier code, allocated at zero cost by the US Department of Defense. For use outside the USA, the equivalent NCAGE code is issued by NATO and other government defence agencies. More information is available at http://www.dlis.dla.mil/cage_welcome.asp.

thermal expansion, exposure to cosmic rays, and corrosion). Furthermore, even for the high-memory RFID tags available today, the memory available may be insufficient to store digital signatures for each record. Even in the future, the amount of data that can be transferred within a reasonable period of time may be very much limited by the maximum data rates provided by the RFID air protocol (up to 640 kbps with EPCglobal UHF Class 1 EPC GenII/ISO 18000-6c, although the data rates may be substantially lower than this in real-world environments).

Clearly the best solution for managing lifecycle data involves a hybrid approach, whereby some essential data may be stored on the part's tag or memory button, although primarily the emphasis should be on recording the information electronically in a standard format and using messaging infrastructure to ensure that the message reaches those organizations that actually use the data in as timely a manner as possible.

Data synchronization

As discussed in the previous section, many of the benefits of deploying auto-ID technologies with aircraft parts are derived primarily from better information sharing within and between organizations, in order to reduce the amount of time spent retrieving information from paper-based filing systems or via exchange of phone calls and faxes. In its simplest form, it would be sufficient to use the auto-ID technology as a carrier for a unique identifier, which can then be used to retrieve information from various databases (including local caches of that information) and to display the relevant information to the mechanic, inspector or other staff, to make the business process more efficient.

Regardless of whether or not any additional data is stored on the RFID tag, it is desirable that relevant information about a particular part should be readily available without delay to the person who is handling the part. Furthermore, it is essential to capture the additional information that they record about their activity on the part, such as details of a maintenance event or observations (such as overheating, leakages, vibrations or emission of fumes), and that this is promptly communicated to the company's own system of records (and potentially onwards to other organizations that have a valid interest in receiving some or all of this information).

Our initial work on the data synchronization theme focused on the technical details of how to ensure that data on the tag and the corresponding data on the back-end database could be synchronized in such a way as to avoid contention (situation in which it is unclear which has the more recent or more authoritative data). The key findings of this approach were that it would be necessary for both the data on the tag and the data on the back-end database to be recorded as journals, so that appropriate merges and (where necessary) rollbacks of the data updates can be performed. This is particularly important where more than one person or organization may be updating the records (either on the tag or on the

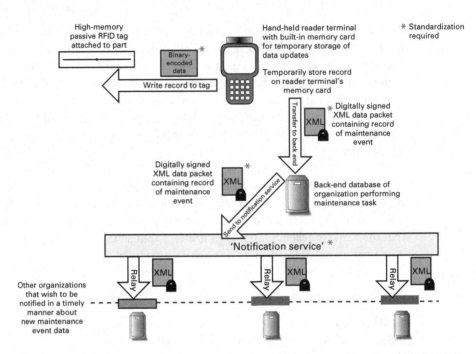

Fig. 10.9   The data synchronization mechanism (from Suzuki and Harrison, 2006, with permission) [10].

back-end database), and especially when there may be delays. For example, when RFID equipment is not available or not functioning properly, it might not be possible to write the update to the tag at the time – but the update should still be recorded and communicated to the back-end database.

As a result of closer involvement with the sponsors of the Aerospace ID Technologies programme and with members of the ATA RFID on Parts project team, we realized that the main priority for data synchronization had to be to ensure the prompt flow of maintenance-event information from the point at which it is generated to the company's own back-end system and, beyond that, to additional organizations that can use this data. (See Fig. 10.9.) We decided to treat the "reverse path" (or two-way synchronization to ensure that the data on the tag is complete) as a lower priority because of limitations on the security and memory capacity of the current generation of tags available – and because the sharing of information via the network enables much greater benefits.

Some aerospace companies take a very modern approach to information sharing and have done their internal data integration, so that they are able to retrieve records from their information systems in response to a query from another organization. However, other organizations are much more hesitant about providing interfaces to their database systems, perhaps because they know that significant internal data integration of their legacy systems is needed as a pre-requisite – and also because many have concerns about ownership of data, privacy/confidentiality of data, liability issues revealed by the data, etc.

One area of common ground is that many aerospace companies today are capable of electronic data interchange, even if it merely takes the form of sending a document or spreadsheet as an e-mail attachment to another organization. An important point to note about this approach (as opposed to the approach of providing a direct web-service interface to a company's database) is that documents can be exchanged – but what action is taken is entirely the decision of the recipient of the document. The recipient can simply store it for future reference – or they can check its authenticity, then choose to extract relevant information from its contents – and then may also choose to store this extracted information within their own databases and take further actions or make decisions as a result of the new information. Recognizing that such a "stepping stone" approach can yield many of the benefits of information sharing, we proposed the architecture (shown in Fig. 10.8) as a mechanism for lifecycle data management and inter-organizational data synchronization.

Other research activities

In addition to work on the five key themes, the Aero ID Programme has initiated activities in the following areas:

- security of RFID deployment in the aerospace sector;
- general approaches for investigating ROI across the aerospace supply chain; and
- the role of RFID in improving airport performance.
- The latter activity, in particular, has led to a number of new spin-off projects, which are currently under way.

10.6 Trials and industrial adoption

Beyond the research, the Aero ID Programme has supported and fostered practical deployments of many of the research outputs within the sponsor organizations. In this section we summarize some of them.

Trials and demonstrators

At the time of writing, the Aerospace ID Technologies Programme had been running for 17 months, during which 12 reports were released to sponsoring organizations, with results from the five program themes. For the purpose of research validation, these findings are being deployed in pilot trials within sponsoring organizations, according to their individual interests and needs. They provide an opportunity to introduce innovation stemming from the program into real industrial processes, and to obtain feedback from front-line users of the automatic identification technologies.

1. *Embraer.* Brazilian aircraft manufacturer Embraer is running a pilot within the logistics chain of its new Embraer 190 aircraft, which will span the first semester of 2007. This trial involves supplier C&D Zodiac, the global leader in aircraft interiors, and its subsidiary C&D Brazil. The object of this experiment is improved tracking and tracing processes, with a focus on visibility across the logistics chain and on timely information about order status and location. A lack of quality and timeliness in the information flow has a direct impact on the level of safety stock built into the chain. The safety stock warrants Embraer's continued operations against any fluctuations in supply and demand but, on the other hand, is rather costly in terms of commitment of working capital. The quality of information has also a direct effect on the decision process bridging production demand and logistics, which in turn impacts operational costs. Embraer will be assessing RFID-enabled logistics processes and their ability to deliver
   - quicker information sharing across the two organizations;
   - smaller inventory levels without compromising production;
   - better efficiency in manpower deployment within logistic processes;
   - collaborative management of the logistics chain; and
   - a smaller gap between market demand and production performance.
2. *Messier-Dowty.* UK-based landing-gear manufacturer Messier-Dowty is on the opposite end of the logistics chain, as a supplier to airframe manufacturers. In some cases contractual agreements require that final assembly of landing gear take place at a plant local to airframe manufacturers in Europe and North America. In that situation a third party comes into play between the landing-gear supplier and the aircraft manufacturer. Messier-Dowty is considering a trial similar to Embraer's in which better visibility of the outbound logistics chain is achieved by the use of automatic identification technologies. As in the previous case, improved traceability and timeliness of information exchanges between the players are the key value drivers being sought.
3. *Aviall.* Aircraft parts distributor Aviall saw an opportunity for internal process improvement arising from the fact that it handles RFID-tagged parts on a daily basis. Its clients that are contractors of the US Department of Defense are required to tag shipments sent to military facilities around the globe. Aviall has been experimenting with these tagged items for better tracking and tracing within its vast warehouses in Dallas, Texas, and reductions of up to 65% in numbers of process steps have been achieved. The company is currently considering extended trials in cooperation with the Aerospace ID Programme and possibly with Boeing, which acquired Aviall in 2006.

In addition to these trials under development, a number of demonstrators are being developed in the program. These tend to be simpler deployments of research results than trials, and are usually working prototypes in the form of software tools. The first example is a prototype mechanism for data synchronization being developed by BT Auto-ID Services. As was explained in Section 5.6, this fulfills requirements from the Boeing 787 aircraft program for secure and reliable synchronization between data stored on tags and Boeing's back-end databases. It draws from program contributions for a unique identifier of aerospace parts, and for standard data fields in the birth and maintenance records of those parts. Additionally, it embodies tag data translation procedures that transform binary data into XML format.

Another area where demonstrators are being considered is sensor integration, as described in Section 10.5. Three Auto-ID Labs – Cambridge, Keio, and the ICU – are cooperating in research into alternative IT architectures that would combine identification data with sensor data. This combination promises to deliver improved visibility of the state of aircraft parts, consequently providing a better understanding about where they stand within their service lives. A prototype representing one or two of the most promising architectures is currently being discussed, and should be deployed before the end of 2007. Several sponsors are likely to become involved in this development.

Furthermore, as part of the program, sponsor Afilias is developing the concept of a notification service that can provide information from sensors associated with a tag (i.e. temperature history, pressure variations, humidity, and so on). Sensor data is fed into a system able to monitor the condition of the part and able to trigger a notification once a threshold is reached. This message is automatically delivered by the notification service to one or more authorized stakeholders, who can then act on that information. The stakeholders can be from different companies, and may be located anywhere in the world.

Standards for the aerospace sector and the Aero ID Programme

The aerospace sector already has a number of standards bodies and industry organizations, such as the Air Transport Association (ATA) and the International Air Transport Association (IATA), the Aerospace Industries Association of America (AIA), SAE, and SITA. In November 2006, a meeting of these standards bodies and others, including EPCglobal Inc. and AIM, was held in order to identify areas of overlap and possible duplication of effort, as well as gaps in the standardization activities. The Auto-ID Lab at Cambridge participated in that meeting, not as a standards body, but because of our close involvement in the technical work, through the Aerospace ID Technologies Programme. The ATA has a long history of developing standards for data formats and messaging about aircraft parts – and their standards are compiled in the various chapters of the ATA Spec 2000 document.

Within the Aerospace ID Technologies Programme, members of the Auto-ID Labs at Cambridge and Keio have worked closely with the ATA's RFID on Parts project team to discuss technical feasibility and security implications for the proposals being developed for unique identifiers and additional data to be recorded on RFID tags for aircraft parts. For the unique identifiers, the program

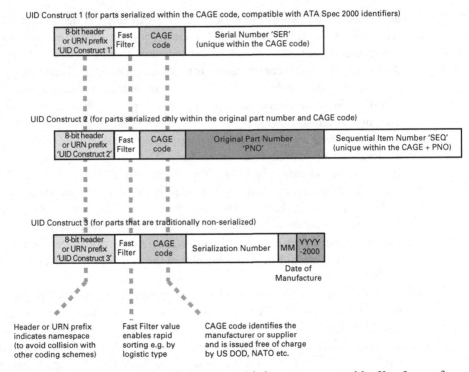

UID Construct 1 (for parts serialized within the CAGE code, compatible with ATA Spec 2000 identifiers)

| 8-bit header or URN prefix 'UID Construct 1' | Fast Filter | CAGE code | Serial Number 'SER' (unique within the CAGE code) |

UID Construct 2 (for parts serialized only within the original part number and CAGE code)

| 8-bit header or URN prefix 'UID Construct 2' | Fast Filter | CAGE code | Original Part Number 'PNO' | Sequential Item Number 'SEQ' (unique within the CAGE + PNO) |

UID Construct 3 (for parts that are traditionally non-serialized)

| 8-bit header or URN prefix 'UID Construct 3' | Fast Filter | CAGE code | Serialization Number | MM | YYYY -2000 |

Date of Manufacture

Header or URN prefix indicates namespace (to avoid collision with other coding schemes)

Fast Filter value enables rapid sorting e.g. by logistic type

CAGE code identifies the manufacturer or supplier and is issued free of charge by US DOD, NATO etc.

Fig. 10.10   Different UID constructs. (Permission to use this image was granted by Ken Jones of ATA.)

has helped the aerospace organizations to gather their user requirements, and then developed a technical proposal for three unique identifier (UID) constructs, to meet their user requirements. The structure of the three identifier constructs is shown in Fig. 10.10. The technical proposal describes how these could be expressed as EPC identifiers, so that the aerospace sector could easily use the readers, tags, middleware, and information services that conform to EPCglobal interface standards, in order to benefit from economies of scale, by using commoditized solutions that have already been developed for other industry sectors (such as the consumer goods/retail sector), where appropriate. The proposal is to be submitted to EPCglobal, and discussions with the US Department of Defense are also going on in order that suppliers to both civil and military aerospace sectors can use a common standard.

The ATA RFID on Parts project team is making progress on the structure of the additional data to be stored on the tag and also the mechanism to synchronize the data to the back-end database or system of record and on to additional organizations. Our research work and participation in the ATA project team meetings aims to provide impartiality, clarity, and further technical guidance in these areas. Figure 10.11 shows the conceptual structure of the data layout for the high-memory tags that are being considered for use with aircraft parts. At the time

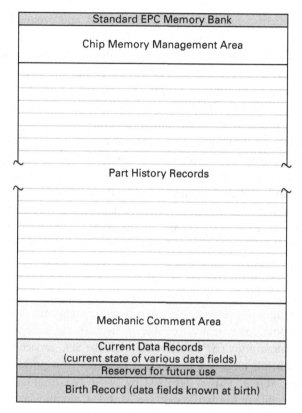

Fig. 10.11   Chip memory allocations (courtesy of Intelleflex).

of writing, the ATA RFID on Parts project team has already determined which data fields from the CSDD should appear in the "birth record" information – and is in the process of deciding which data fields should appear in maintenance event records.

Sponsor developments

Aerospace ID Programme sponsors are encouraged to integrate research results into their own product offerings. IT systems integrators and service providers have a vested interest in improving their business with innovation stemming from our contract R&D. The following are examples of new solutions and approaches implemented at the current stage of the programme.

1. *SITA/VI Agents/Afilias partnership.* SITA is an IT applications and communications services provider for the air transport industry. The company teamed up with Aerospace ID Programme co-sponsors VI Agents and Afilias to launch their joint service in January of 2007. In simple terms, the service first captures auto-ID-generated data from tagged assets, baggage, or aircraft parts. It then

feeds the data to client-defined business process rules, and eventually notifies any pertinent stakeholders if relevant events are detected by these rules. VI Agents provides the EPCIS-compatible software that embodies the business processes for open-loop and closed-loop applications, Afilias developed the discovery service that provides links across multiple organizations, and SITA runs the information technology infrastructure as well as managing the customer front end. Detailed information can be found at http://sita.autoid.aero/yacs/.

2. *BT Auto-ID Services.* Building on its extensive experience in the retail, automotive, and construction sectors, BT Auto-ID Services announced AeroNet in October of 2006. This offering is a fully managed service that not only delivers auto-ID-enabled processes to the civil aviation sector, but also provides compliance to American and British defense requirements. BT Auto-ID Services is also working on its version of an EPC-compatible discovery service that should be deployed across its offerings to several industrial sectors besides aerospace.

3. *T-Systems.* The German IT services provider T-Systems also began offering a managed service in 2006, which has been fine-tuned to meet aerospace requirements. It builds on experience gained from client engagements with Airbus and Frankfurt Airport, to name just a couple of clients in the aerospace sector.

4. *System Planning Corporation (SPC).* GlobalTrak is the SPC service for tracking location and condition of containers across multiple transportation modes (i.e. trucks, trains, ships). This is achieved by two-way communication with cargo via either satellite or Global System for Mobile Communications cellular networks. Real-time sensor monitoring and alert notification are available and driven by client-defined rules. As a result of involvement in the Aerospace ID Programme, SPC further developed GlobalTrak hardware into Class 5 tags. Class 5 RFID tags are able to communicate with other tags, having their own power source and embedded reader. This means that it is now possible to track and trace not only large shipping containers fitted with GlobalTrak hardware, but also individual boxes or items within the containers themselves.

5. *AEROid Ltd.* This UK-based RFID systems integrator has been able to customize an entire line of offerings directed at the commercial and defense aviation sectors as a result of participation in the program. Examples of these innovative products are tracking systems for spare parts and line replaceable units, for ground support equipment at airports, and for the catering trolleys that take food on board aircraft. It is also a player in industrial applications such as RFID-enabled kanban for aircraft assembly lines.

10.7   Summary

The Aerospace ID Technologies Programme has been established to support ID system adoption in the aerospace sector through industrially relevant research. Through a series of research themes the program has developed a diverse set of outputs, many of which are being put to immediate use by sponsors and more broadly within the industry. By collaborating tightly with sponsors the point at which the research stops and the development and adoption processes starts is blurred and as the program comes to an end in mid 2007 a number of activities will continue:

- advising and providing recommendations for standards development processes,
- development and trialling of aero-industry-specific solutions by program sponsors,
- turning program deliverables into openly accessible tools,
- running of an ongoing forum for ensuring that industry needs for development can be captured and assessed, and
- focusing studies into key areas identified as requiring ongoing research.

10.8 Bibliography

[1] **Hall, D. L.**, *Mathematical Techniques in Multi-Sensor Data Fusion* (Artech House, Norwood, MA, 1992).

[2] **Bosse, E.**, **Roy, J.**, and **Grenier, D.**, "Data Fusion Concepts Applied to a Suite of Dissimilar Sensors," in *Canadian Conference on Electrical and Computer Engineering*, vol. 2, pp. 692–695 (1996).

[3] **Grossmann, P.**, "Multisensor Data Fusion," *The GEC Journal of Technology*, 5:27–37 (1998).

[4] **Harrison, M.**, and **Shaw, A.**, *Electronic Pedigree and Authentication Issues for Aerospace Part Tracking* (Auto-ID Lab, Cambridge, 2006).

[5] **Thorne, A.**, **McFarlane, D.**, **Le Goff, K.**, and **Parlikad, A.**, *Scoping of ID Application Matching* (Auto-ID Lab, Cambridge).

[6] **Brusey, J.**, and **Thorne, A.**, *Aero-ID Sensor Integration: Scope of Work* (Auto-ID Lab, Cambridge, 2006).

[7] **Kelepouris, T.**, **Theodorou, S.**, **McFarlane, D.**, **Thorne, A.**, **Harrison, M.**, *Track and Trace Requirements Scoping* (Auto-ID Lab, Cambridge, 2006).

[8] **Harrison, M.**, and **Parlikad, A. K.**, *Lifecycle ID and Lifecycle Data Management* (Auto-ID Lab, Cambridge, 2006).

[9] **Sharp, E.**, *Technology Selection for Identification Applications* (Auto-ID Lab, Cambridge, 2006).

[10] **Suzuki, S.**, and **Harrison, M.**, *Data Synchronization Specification* (Auto-ID Lab, Japan, and Auto-ID Lab, Cambridge, 2006).

[11] **Kelepouris, T.**, **Baynham, T.**, and **McFarlane, D.**, *Track and Trace Case Studies Report* (Auto-ID Lab, Cambridge, 2006).

[12] **Patkai, B.**, and **McFarlane, D.**, *RFID-based Sensor Integration in Aerospace* (Auto-ID Lab, Cambridge, 2007).

[13] **Kelepouris, T.**, **Bloch da Silva, S.**, and **McFarlane, D.**, *Automatic ID Systems: Enablers for Track and Trace Performance* (Auto-ID Lab, Cambridge, 2007).

[14] **Harrison, M.**, *EPC Identifiers for Aerospace* (Auto-ID Lab, Cambridge, 2007).

[15] **Porad, K.**, and **Heitmann, J.**, *Boeing-Airbus Global RFID Forum* (2005).

11   The cold chain

J. P. Emond

The cold chain is a concept resulting from the field of the transformation and distribution of temperature-sensitive products. It refers to the need to control the temperature in order to prevent the growth of micro-organisms and deterioration of biological products during processing, storage, and distribution. The cold chain includes all segments of the transfer of food from the producer to the consumer. Each stage crossed by a temperature-sensitive product is related to the preceding one and has an impact on the following one. Thus, when a link of this "cold chain" fails, it inevitably results in a loss of quality and revenue, and, in many cases, leads to spoilage.

The cold chain concept can also be applied to many other industries (pharmaceuticals, dangerous goods, electronics, artifacts, etc.) that require the transport of products needing to be kept within a precise temperature range, in particular at temperatures close to 0 °C, or even below [1]. Owing to very strict rules from government agencies the expression "cold chain" has even become one of the key sentences at the heart of current concerns in the pharmaceutical field and in biotechnology [2].

11.1   The food industry

Temperature is the characteristic of the post-harvest environment that has the greatest impact on the storage life of perishable food products. In some regions of the globe, especially in tropical and subtropical regions, post-harvest losses of horticultural crops are estimated to be more than 50% of the production due to poor post-harvest handling techniques such as bad temperature management. For example, in India a large quantity of onions is lost between the field and the consumer due to lack of adequate post-harvest handling procedures [3]. Good temperature management is, in fact, the most important and simplest procedure for delaying the deterioration of food products. In addition, storage at the optimum temperature retards aging of fruit and vegetables, softening, and changes in texture and color, as well as slowing undesirable metabolic changes, moisture loss, and loss of edibility due to invasion by pathogens [4]. Temperature is also a factor that can be easily and promptly controlled.

RFID Technology and Applications, eds. Stephen B. Miles, Sanjay E. Sharma, and John R. Williams. Published by Cambridge University Press. © Cambridge University Press 2008.

Low temperatures during storage of fresh food products depress both physio-logical processes and activities of micro-organisms capable of causing product spoilage. Even worse than product spoilage is that pathogenic flora or toxins may cause food-borne diseases. Controlling the cold chain is essential in order to prevent the multiplication of micro-organisms. Billard reported that, under optimal conditions, bacteria divide every 20 minutes [5]. A typical transportation duration of 8 hours under these optimal conditions will allow a bacterium to generate over 16 million descendants. Low temperature is also critical for keeping food free from the pathogenic flora or toxins responsible for many outbreaks of food borne diseases around the world. The US Department of Agriculture (USDA) has estimated that medical costs and productivity losses associated with the main pathogens in food range between $6.5 billion and $34.9 billion annually. This confirmed that controlling the cold chain is essential as a means to reduce microbiological risks.

The food industry has developed a system called hazard analysis and critical control points (HACCP) that is a methodical approach to food safety that addresses physical, chemical, and biological hazards. This system involves looking at all stages of food production and its distribution in terms of deploying means of prevention of contamination rather than simply inspecting the finished product. Temperature management is associated with one of the most important potentials for food safety hazards and that is why it is known as a critical control point, which requires a strict recoding of any deviation from the specifications. The US Food and Drug Administration (FDA) clearly stipulates that all raw materials and products should be stored under sanitary conditions and under the proper environmental conditions such as temperature and humidity in order to assure their safety and wholesomeness [6]. The HACCP system has been proven to be a successful method to assure quality of temperature-sensitive products and is widely used in the food industry, including in the food service and restaurant industries.

Even though food safety is the first priority in the food industry, gaining markets is the priority on the business side. In the USA alone, the retail food industry generates around $950 billion in sales and employs approximately 3.5 million people. More than 50% of these sales each year are generated by perish-able food products such as produce, meat, dairy, fish, and baked goods. In order to maintain adequate revenues, retail stores must rely on volume since the average profit is only about 1.4%. The key factors for this industry include limiting losses, ensuring safety, and always innovating in order to keep or gain market share. One of the best ways to keep or gain market share is by presenting a perfect produce section. This positive image gives customers a better perception of the quality of the business. In the minds of many customers, if the store provides high-quality fresh produce, it is probably maintaining equally high quality for the other products in the store. However, the price to be paid for keeping this "image of freshness" is the need to have an inventory turnover of almost 50% per day. This is the highest percentage in a retail store, followed by the meat and fish sections.

This explains why the average loss of potential revenue from produce for a supermarket is about $200,000 per year [7]. About 85% of these losses are due to deficiency in the cold chain [8].

In general, the lower the storage temperature within the limits acceptable for each type of commodity (above their freezing point), the longer the storage life. Results of several studies have demonstrated that maintaining an optimum temperature during storage and transport is crucial for keeping vegetable quality. Fruits and vegetables are, in fact, highly perishable products, and losses due to inadequate temperature management are found to be mainly due to water loss and decay [9]. Some fruit and vegetables of temperate, subtropical, and tropical origin such as papayas, mangoes, tomatoes, cucumbers, bell peppers, and others must be kept at temperatures between 10 °C and 18 °C. Upon exposure to lower temperatures and according to their respective sensitivities to cold storage, these products will inevitably develop symptoms of chilling injury. These symptoms might not be apparent and will not show up immediately, but will lead to permanent and irreversible physiological damage. So, there are real opportunities to improve the cold chain from the growers of produce to the shelves of a retail store.

11.2   Pharmaceuticals

At the beginning of the 1980s, much effort was deployed to ensure an uninterrupted cold chain of pharmaceutical products such as vaccines by using optimized equipment such as temperature recorders and insulated packages and providing better training to the various parties involved in the distribution system. However, it is only recently that significant improvements were set into motion to increase cold chain control and maintenance for temperature-sensitive pharmaceutical products. Even though significant improvements have been observed, Brandau estimated that half of all vaccines distributed are lost or wasted [10]. Although many failures in the distribution systems of third-world countries contributed to the rate of loss of vaccines, many of these losses were also observed in industrialized countries.

Result from studies [11], [12], [13], and [14] showed that, even though refrigeration is often taken for granted in industrialized countries, handling errors resulting in breaks in the cold chain occur much more frequently than had initially been estimated. This is a critical issue since the inactivation of a vaccine may become apparent only after patients administered the product acquire the disease it was designed to prevent [15].

In order to regularize matters and to ensure an optimal quality of the vaccines, the World Health Organization, as well as other government agencies such as the FDA and Health Canada, require a temperature of 2 °C to 8 °C for the transport and storage of the majority of vaccines. That regulation excludes those having to be kept frozen, such as the oral vaccine against poliomyelitis [16] and the vaccine against smallpox in dry form [17], which must be stored at –20 °C. The importance of proper temperature management can be clearly illustrated using the example of the oral polio vaccine. This vaccine is stable for 6–12 months at the recommended storage temperature whereas it becomes very unstable at 41 °C, losing 50% of its activity in just 1 day. When exposed to a temperature of 50 °C it crosses the lower limit of acceptable activity within 3 hours [16] [18]. Any break in the cold chain can lead to a decrease of the therapeutic efficiency of the vaccines and an increased rate of side effects, and hence a decrease of the confidence of the population and health professionals in regard to vaccination, without counting the increase in cost of immunization programs [19].

It is the responsibility of pharmaceutical companies or wholesalers to assure that every shipment regulated by government agencies such as the FDA remains within the temperature ranges required throughout the whole distribution process. Any deviations must be reported and may result in the load having to be recalled and destroyed. Most temperature-sensitive pharmaceutical products are shipped in insulated containers designed to provide thermal protection. These containers are usually "validated," meaning that they go through rigorous standardized testing showing that they can maintain the stipulated temperature ranges during distribution under extreme environmental conditions. Using validated containers does not require the use of temperature-recording devices for most shipments. According to the FDA Code of Federal Regulations Title 21, Section 211.94(b) [20], "Container closure systems shall provide adequate protection against foreseeable external factors in storage and use that can cause deterioration or contamination of the drug product." However, unless the shipper can provide data that show the validation process, all shipments must be monitored. To help in the design of "validated" containers, the Parental Drug Association's Technical Report No. 39 [21] provides guidance to the pharmaceutical industry and regulators on the essential principles and practices for shipment of products that require controlled temperatures during transit, while providing a design approach to the development of specialized packages and systems that will protect temperature-sensitive products during transportation.

Validated containers are widely used in the pharmaceutical industry as well as in blood banks around the world. This growing technique is not intended to use temperature-monitoring devices and would not contribute in the short term to the RFID market. However, new potential applications discussed further on in this chapter may contribute to the adoption of RFID temperature-monitoring devices.

11.3   Types of temperature-tracking technologies

Tracking of temperature is used when temperature-sensitive products are transported and potentially exposed to harmful environmental conditions. For some users, just knowing that the shipment was kept within a specific range (high and low limits) is enough to confidently use the product. In some other cases (particularly pharmaceutical products), knowing that the product exceeded a specific temperature for a certain period of time is adequate for deciding whether to reject or accept a shipment. There are simple indicators of temperature such as a vaccine vial monitor (VVM) manufactured in a thermo-sensitive material that one affixes on to the label or the stopper of a vaccine vial; it records the exposure to heat over time [18]. The combined effects of time and temperature make the VVM change color gradually and in an irreversible way. Other temperature indicators based on the same principle, such as the control sheets of the cold chain, FreezeWatch™ and MonitorMark™ (Time Temperature Indicators, 3M, St. Paul, MN), are available and can be inserted into the shipment.

For many years, temperature recording was done using chart recorders that were retrieved upon arrival and read to detect any temperature deviations during transit. This technology was abandoned for digital temperature loggers featuring read/write memory, a digital thermometer, a real-time clock, programmable high- and low-temperature thresholds, and light-emitting-diode lights that indicate whether temperature requirements have been maintained. These digital loggers such as TempTale (Sensitech, Inc.) are widely used for tracking shipments of perishable products but have the disadvantage of requiring to be connected physically to a computer to download the data. This process makes the acquisition of a complete temperature history of a shipment a very manual operation. Furthermore, it is almost impossible to read the data during transit.

Having the possibility to read temperature during the distribution process adds a lot of value to a temperature-monitoring solution. Temperature information gathered by a typical digital logger usually becomes available only at destination and this reduces significantly its helpfulness when users are facing the "facts" at the end of the shipping process. The real-time monitoring of RF technology makes it a major component in a successful temperature-monitoring program.

As covered in previous chapters, there are "passive," "semi-active," and "active" RF tag technologies. All of them can offer the possibility of providing temperature readings during the distribution process. Only the "semi-active" and "active" technologies can provide the possibility to store data. Nevertheless, the use of "passive" RFID temperature tag technology like that of Gentag (Washington, DC) can allow users to monitor temperature wirelessly and can feed a mainframe computer with data read at a programmed acquisition frequency. This technology is the cheapest way of monitoring temperature using RF technology. The downside is that it does not allow storage of data during the transit time.

"Semi-active" RFID temperature tags have been the focus of several tests in recent years [22]. The major component of a successful trial with this technology relies on the frequency used by these devices. Many HF "semi-active" tags working at 13.56 MHz have proved the concept of wireless temperature monitoring, such as ThermAssureRF (Evidencia, Memphis, TN). One of the main advantages is the ability to read the tag through liquid, which is the main component of many food and pharmaceutical products. These tags are very light and are fully waterproof, which can be a major advantage for products like fresh fish. The major weakness of this frequency is the short transmission range (<1 m). The other well-known frequency for this family of "semi-active" tags is 2.4 GHz, as used for the Alien battery tag. Even though such tags have a longer transmission range (3–5 m) their ability to transmit in a "humid" environment is weaker than that of tags operating at 13.56 MHz.

The major advantage of a semi-active (class 3) RF tag is that it can safely be flown on any aircraft, including in loads carried on commercial flights; this use has since been approved by the Federal Aviation Administration (FAA). This feature is very important, considering the large amount of valuable food and pharmaceutical products transported by aircraft around the world. In fact, about 20% of all cargo shipped by air is perishable and pharmaceutical products [23]. This does not include all the pharmaceutical and blood products air-shipped as overnight parcels.

Active RFID tags use their own internal power source, like semi-active tags, but also usually generate a continuous outgoing signal. Active temperature RFID tags like the Identec 915-MHz tag (Houghton-le-Spring, UK) give the possibility to read the tag from a distance of at least 30 m. Whereas 915-MHz passive RFID tags have proven very challenging to read in the presence of water, the situation is slightly better with the active version [24]. Long read distances may seem a good advantage over other types of RFID temperature tags but can also be a disadvantage in a distribution environment. The possibility of reading a tag that is not in the shipment being interrogated may provoke a lot of confusion at a distribution center. Active RFID temperature tags are all Class 4 and not yet approved by the FAA. As discussed previously, the amount of perishables and pharmaceutical products transiting by air is very significant and such technology would not be adequate. All these RFID temperature tags must be fully encoded with a valid EPC code (or some other code) so that they can be identified with their own ID. An individual ID allows a temperature history to be associated with a specific load (examples of significant variability of temperature in the same shipment are presented later in this chapter).

Finally, new features that are of great use for temperature tracking of food and pharmaceutical shipments are now available on the market, such as stamping and proof of read–write interaction with the tag (Evidencia, Memphis, TN). This feature allows users to see who has read the tags and when. In a claim situation this feature may be critical. The possibility of having a temperature probe is also a great advantage, as will be seen later in this chapter.

11.4   Challenges associated with RFID temperature-tracking technologies

The food industry

Food products are known to present challenges for RFID technology. The amount of water in food such as fresh produce as well as their packaging materials can interfere significantly with radio signals. Packages are usually designed to keep the freshness of food by preventing interactions with gases, pathogens or humidity that are known to shorten shelf life. For example, oxygen is liable to decrease the shelf life of many processed foods. In order to achieve these protections, packaging must use materials called "barriers." These barriers have the property of being able to stop any transfers of aroma, gases, water or light. Most of them include aluminum foil or Mylar, which does not let radio signals pass through. The main packaging materials identified as not being friendly for RF signals are metal or aluminum cans, aluminum foil, Mylar film (potato chips bags), and multilayer paper packages (e.g. juice or UHT milk boxes).

More than the problem of reading the signal due to the presence of water or packaging materials, the real issue involved in using RFID temperature tags in monitoring shipments is mostly that of knowing "which" temperature the tag is giving. Owing to the difficulty of reading a radio signal at some frequencies in the middle of a load of food product, many users will place the RFID temperature tag at an outside location on the load. This practice may provide wrong information about the "real" temperature of a load. As shown in Fig. 11.1, the temperature measured on the outside of a pallet of radishes was completely different than the temperature of the product. This discrepancy can lead to the rejection of a load when no real problem was encountered.

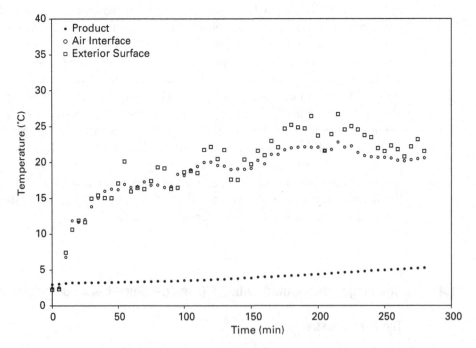

Fig. 11.1   Temperature measurement of a pallet of radishes exposed to the Sun on a loading dock (Pelletier, IFAS 2004 internal communication, with permission) [25].

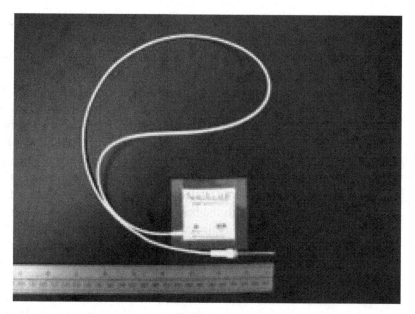

Fig. 11.2   An RFID temperature tag operating at 13.56 MHz with a temperature probe.

Unless the RFID temperature tag uses a frequency able to be read at different locations in a load, RFID technology will have to rely on algorithms to convert the "measured temperature" into the "real temperature" of a shipment. Evidencia (ThermAssureRF) is proposing a solution with a temperature probe attached to the RFID temperature tag. This HF tag has a 0.4-m temperature probe (Fig. 11.2) that can reach any location in a pallet or load. Even though the tag is attached to the outside of the load, the probe can provide the temperature of the product.

The pharmaceutical industry

RFID challenges with pharmaceutical and blood products are very similar to those discussed in the section on food products. These products, particularly vaccines and blood, contain a very high amount of water. Their packaging systems are more complex and rarely use RF-friendly materials. The two levels of packaging used during a typical shipment of pharmaceutical product present different challenges for monitoring temperature using RFID technology. The first level, known as primary packaging, is in direct contact with the product. These packages are made of glass or plastic and do not offer a great contact from which to measure their temperatures. So the temperature measured will usually be that of the surrounding air, which is likely to change more rapidly than that of the product itself. The second level presents the highest level of difficulty for RF technologies. This packaging is designed to thermally protect the products against temperature abuses. The insulating materials used for these packages are mainly expanded polystyrene, urethane, and vacuum-insulated panels (VIPs). All of them let radio signals penetrate the package, except for VIPs, which constitute a complete "barrier" to RF due to the use of Mylar for external protection. Another component of this protective package is the temperature stabilizer (TS). A TS provides cold for shipments during summer and serves as thermal mass during winter shipments. Most of them are mainly composed of water and will be placed between the products and the protective package, which will significantly affect RF technologies using UHF frequencies such as 915 MHz. Therefore the location of the RFID temperature tag is critical when one wants to measure the shipment temperature but also to read the tag without tampering with the packaging system.

There are no specific recommendations about the best location to place a temperature recorder inside an insulated container. The only recommendation is that they should be located with the best possible judgment regarding the type of packaging. Of course, it is best to place a heat monitor at the point of greatest heat exposure (usually near a side, away from cold packs); likewise for a freezing-temperature monitor, which must be placed with vaccines near cold packs in summer and also near a side but away from cold packs in winter. In the National Guidelines for Vaccines Storage and Transportation written by Health Canada, the only recommendation about temperature indicators is that "Manufacturer and central pharmacies in Canada should place temperature monitors in their shipments of vaccines. If justified by the amount of vaccine shipped, monitoring devices that record shipping conditions should be used." [26]. The FDA wants assurance that any indicator is placed in a sensitive location, where it would be likely to detect an out-of-range condition [2].

Dea published data about the temperature profiles of vaccines transported in insulated boxes [27]. The results shown in Fig. 11.3 demonstrate the significant variability of temperature measured at different locations in a package. The location of the RFID temperature tag must be well studied prior to being used in regular shipments since any deviations outside the stipulated temperature ranges must be reported and explained.

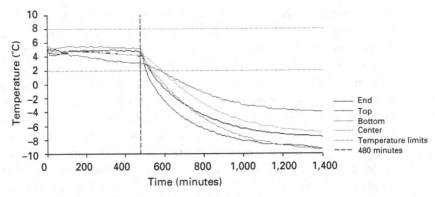

Fig. 11.3   Temperature profiles at different locations inside an insulated box of vaccines exposed at 4 °C for 480 minutes and transferred to 30 °C for 960 minutes (Dea, Masters Thesis, Lavalle University, 2005, with permission) [27].

In the pharmaceutical industry, temperature-recording devices must be calibrated with a standard method using a certified temperature sensor every 6–12 months. RFID temperature-sensor manufacturers such as Evidencia provide RFID temperature monitors calibrated in 99.2% of all units at ±0.2 °C for pharmaceutical applications.

11.5 Potential applications in "semi- and real-time" cold chain management

The food industry

The use of RFID temperature monitoring offers a new way of managing the cold chain in the food industry. Rather than reacting to the fact at the distribution center, food suppliers and retailers can rely on partial or full visibility of the condition of a shipment prior to its arrival at its destination. Shipments of food products having cold chain problems may be redirected to nearby distribution centers or returned to their origins rather than having them continue their trips to the final destination. This will enable retailers and suppliers to plan the movement of perishable products more efficiently, thus reducing waste and providing better-quality products to consumers. Pelletier *et al.* [28] demonstrated that knowing the full temperature history from the field to the distribution center can make a difference of as much as $48,000 in loss of revenue per trailer of strawberries for a retailer.

This full visibility can allow retailers, food services and restaurant chains to manage their inventories by "First to expire, first out" rather than "first in, first out" by combining real-time data with predictive shelf-life models like the ones developed by Nunes *et al.* [29]. These models can integrate the quality attributes selected by the users and predict which loads should be sent to specific stores or restaurants in order to maximize inventories and quality with visible results (Fig. 11.4).

Monitoring temperature in real time for perishable products such as fresh produce presents a great opportunity for the food industry. However, since RFID smart tags can handle other kinds of sensors, humidity sensors would complement very well a cold chain management program for fresh produce. Water loss from fresh produce is the first symptom perceived by consumers. Monitoring humidity during transit can reduce substantially the waste at the retail level.

The pharmaceutical and blood industry

In most shipments prepared for the pharmaceutical and blood industry, temperature is not always monitored. Validated shipping containers are used for this purpose and are accepted by agencies like the FDA if validation data are available and temperature is monitored occasionally. However, the validation process for these containers requires many tests and potential scenarios to "prove" that the

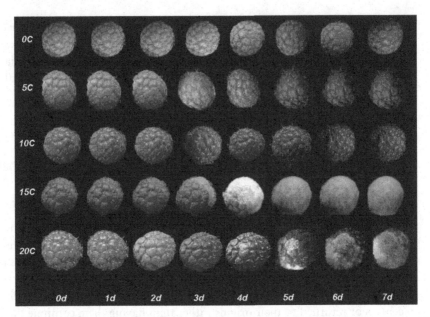

Fig. 11.4   Effects of temperature on quality of raspberries over time (Nunes *et al.*, 2004, with permission) [29].

packaging system can maintain the products within a required temperature range. The validated shipping container is good only for a specific package configuration and cannot be used if one component has been changed (even by a minor change). The shipment of pharmaceutical and blood products can be done in non-validated containers if the temperature is monitored throughout the whole distribution process. Since the validation process is very costly, monitoring temperature can be a very good solution for many companies.

One of the weaknesses of a validated container is the absence of tools to diagnose failures in the distribution system as well as confirmation that the product is still good even though shipping took longer than the duration used in the "validation process." RFID temperature tracking offers the possibility of having full visibility of the distribution system, making it possible to indicate immediately to customers whether the products were kept within the temperature range required by the governing agencies. Furthermore, it allows shippers and distributors to identify weaknesses or failures of equipment prior to having a large amount of products exposed to temperature abuse. It allows users and shippers to verify the temperature of a product prior to opening the container, thus saving time and discussions about claims.

Finally, blood banks can benefit significantly from real-time visibility of the cold chain. Most of them use wet ice to keep products within a specific temperature range. Knowing the actual temperature in a container may dictate re-icing a shipment prior to doing another leg of distribution and thus prevent vital blood units from being wasted.

11.6   References

[1] **Idiquin, M.** *et al.* "La distribution des médicaments en chaîne du froid: carnets pratiques," *S.T.P. Pharma Pratiques*, 7(3):195–212 (1997).

[2] **Walker, L. A.**, "Understanding the Evolution of Cold Chain Management," *American Pharmaceutical Outsourcing*, 2–4 (2002).

[3] **Desai, B. B.**, and **Salunkhe, D. K.**, "Fruits and Vegetables," in *Foods of Plant Origin. Production, Technology, and Human Nutrition*, ed. D. K. Salunkhe and **S. S.** Deshpande, pp. 301–355 (Avi Book, New York, 1991).

[4] **Nunes, M. C. N.**, **Emond, J. P.**, and **Brecht, J. K.**, "Temperature Abuse during Ground and In-flight Handling Operations Affects Quality of Snap Beans, *HortScience*, 36:510 (2001).

[5] **Billiard, F.**, "La logistique du froid dans le commerce de détail," *Revue Générale du Froid*, 967(10):38–44 (1996).

[6] **National Advisory Committee on Microbiological Criteria for Foods**, *Hazard Analysis and Critical Control Point Principles and Application Guidelines* (US Food and Drug Administration, US Department of Agriculture, Washington, DC, 1997) (http://www. cfsan.fda.gov/~comm/nacmcfp.html).

[7] **Villeneuve, S.**, **Nunes, C.**, **Dea, S.**, **Emond, J. P.**, and **Mercier, F.**, *Comptoirs réfrigérés pour fruits et légumes frais, Report to Provigo Distribution Inc.*, 83 pages (1998).

[8] **Villeneuve, S.**, **Emond, J. P.**, **Mercier, F.**, and **Nunes, M. C. N.** "Analyse de la température de l'air dans un comptoir réfrigéré," *Revue Générale du Froid* 1025:17–21 (2002).

[9] **Van den Berg, L.**, "The Role of Humidity, Temperature, and Atmospheric Composition in Maintaining Vegetable Quality during Storage," in *Supermarket Facts. Industry Overview 2002*, ed. R. Teranishi, pp. 95–107 (Food Marketing Institute, 2003) (http:// www.fmi.org/facts_figs/superfact.htm.)

[10] **Brandau, D. T.**, **Jones, L. S.**, **Wiethoff, C. M.**, **Rexroad, J.**, and **Middaugh, C. R.**, "Thermal Stability of Vaccines, *Journal of Pharmaceutical Sciences*, 92(2):218–231 (2003).

[11] **Briggs, H.**, and **Ilett, S.**, "Weak Link in Vaccine Cold Chain," *British Medical Journal*, 306:557–558 (1993).

[12] **Hunter, S.**, "Storage of Vaccines in General Practice," *British Medical Journal*, 299:661–662 (1989).

[13] **Haworth, R. B.**, **Stirzaker, L.**, **Wilkes, S.**, and **Battersby, A.**, "Is the Cold Chain Maintained in General Practice?," *British Medical Journal*, 307:242–244 (1993).

[14] **Yogini, T.**, and **Woods, S.**, "Storage of Vaccines in the Community: Weak Link in the Cold Chain?," *British Medical Journal*, 304:756–758 (1992).

[15] **Casto, D. T.**, and **Brunell, P. A.**, "Safe Handling of Vaccines," *Pediatrics*, 87(1): 108–112 (1991).

[16] **Jaspal, S.**, **Kanta Gupta, C.**, **Sharma, B.**, **Singh, H.**, "Stability of Oral Polio Vaccine at Different Temperatures," *Vaccine*, 6:12–13 (1988).

[17] **Pipkin, P. A.**, and **Minor, P. D.**, "Studies on the Loss of Infectivity of Live Type 3 Polio Vaccine on Storage," *Biologicals*, 26:17–23 (1998).

[18] **Department of Vaccines and Other Biologicals**, *Temperature Monitors for Vaccines and the Cold Chain* (World Health Organization, Geneva, 1999) (http://www.who.int/ vaccines-documents/DocsPDF/www9804.pdf).

[19] **Alain, L.**, *La qualité des vaccins, j'y tiens! Guide des normes et pratiques de gestions des vaccins à l'attention des vaccinateurs – médecins et infirmières* (Ministère de la Santé et des Services Sociaux, Gouvernement du Québec, 1996).

[20] **FDA**, "Biological Drug Products Inspection of Biological Drug Products (CBER)," in *Compliance Program Guidance Manual* (FDA, Washington, DC, 2008) (http://www.fda.gov/cber/cpg/7345848.pdf).

[21] **PDA**, *Guidance for Temperature Controlled Medicinal Products: Maintaining the Quality of Temperature-Sensitive Medicinal Products through the Transportation Environment* (PDA, 2007) (http://www.pda.org/webmodules/webarticles/templates/new_about_newsroom1.aspx?articleid=1034 & zoneid=86).

[22] **Roberti, M.**, "Sensing New RFID Opportunities – Companies Are Beginning to Examine the Benefits of Semi-active RFID Tags that Can Support Onboard Sensors, to Monitor the Conditions of Products and Assets," *RFID Journal* (2006) (http://www.rfidjournal.com/article/articleview/2081/).

[23] **Laurin, E., Cecilia, M., Nunes, N., Emond, J. P.**, and **Brecht, J. K.**, "Quality of Strawberries after Simulated Air Freight Conditions," in *ISHS Acta Horticulturae 604: International Conference on Quality in Chains. An Integrated View on Fruit and Vegetable Quality*.

[24] **Emond, J. P.**, "HF versus UHF in Pharmaceutical Supply Chains," in *5th RFID Academic Convocation – Pre-conference to RFID Live!* (Orlando, FL, 2007) (http://autoidlabs.mit.edu/CS/blogs/presentations/atom.aspx).

[25] **Pelletier, W.**, "Heat Transfer within a Pallet of Radishes Exposed to Solar Radiation," internal communication (Center for Food Distribution and Retailing, Institute of Food and Agricultural Sciences, University of Florida, 2004).

[26] **Health Canada**, *Lignes directrices nationales pour la conservation et le transport des vaccins. Relevé des maladies transmissibles au Canada*, pp. 21–11, 93–104 (Health Canada, 1995).

[27] **Dea, S.**, *Thermal Behavior of Shipping Containers for Temperature Sensitive Pharmaceutical Products*, M.Sc. dissertation (Université Laval de Québec, 2005).

[28] **Pelletier, W., Emond, J. P.**, and **Chau, K. V.**, *Effects of Post Harvest Temperature Regimes on Quality of Strawberries* (University of Florida, 2006).

[29] **Nunes, M. C. N., Emond, J.-P.**, and **Brecht, J. K.**, "Predicting Shelf Life and Quality of Raspberries under Different Storage Temperatures," *Acta Horticulturae* 628:599–606 (2003).

12 The application of RFID as anti-counterfeiting technique: issues and opportunities

Thorsten Staake, Florian Michahelles, and Elgar Fleisch

In industries including aerospace, pharmaceuticals, and perishables, as covered in the preceding chapters, counterfeit trade has developed into a severe problem. While established security features such as holograms, micro printings, and chemical markers do not seem to efficiently avert trade in illicit imitation products, RFID technology, with its potential to automate product authentications, may become a powerful tool to enhance brand and product protection. The following contribution aims at structuring the requirement definition for such a system by providing a non-formal attack model, and outlines several principal approaches to RFID-based solutions.

12.1 Counterfeit trade and implications for affected enterprises

Intangible assets constitute a considerable share of many companies' equity. They are often the result of extensive investments in research and development, careful brand management, and a consistent pledge to deliver high quality and exclusiveness. However, the growth of markets in Asia, where these intangible assets are difficult to protect, the trend in favor of dismantling border controls to ease the flow of international trade, and the increasing interaction of organizations in disparate locations require new measures to protect these assets and safeguard companies from unfair competition. Especially product counterfeiting, the unauthorized manufacturing of articles that mimic certain characteristics of genuine goods and may thus pass off as products of licit companies, has developed into a threat to consumers and brand owners alike.

Counterfeit trade appears to affect a wide range of industries. Alongside the traditionally forged items such as designer clothing, branded sportswear, fashion

RFID Technology and Applications, eds. Stephen B. Miles, Sanjay E. Sharma, and John R. Williams. Published by Cambridge University Press. © Cambridge University Press 2008.

accessories, tobacco products, and digital media, customs statistics show a considerable growth of fakes among consumer products as well as among semi-finished and industrial goods including foodstuffs, pharmaceuticals, fast-moving consumer goods, electrical equipment, mechanical spare parts, and electronic components [1].

The implications are numerous and wide-ranging. Counterfeiting undermines the beneficial effects of intellectual property rights and the concept of brands as it affects the return on investment in research, development, and company goodwill. Producers of reputable products are deterred from investing within a national economy as long as their intellectual property is at risk. National tax income is reduced since fake goods are largely manufactured by unregistered organizations. Social implications result from the above-mentioned costs: the society pays for the distorted competition, eventually leading to less-innovative products and a less secure environment as earnings from counterfeiting are often used to finance other illegal activities [2]. However, for selected emerging markets, the phenomenon also constitutes a significant source of income and an important element of their industrial learning and knowledge-transfer strategy. As a consequence, not all governments determinedly prosecute counterfeiters, which often renders legal measures to eradicate the source of illicit goods ineffective.

For companies, counterfeit trade can lead to

(i)   a direct loss of revenue since counterfeit products, at least partly, replace genuine articles;
(ii) a reduction of the companies' goodwill as the presence of imitation products can diminish the exclusiveness of affected brands and the perceived quality of a product; and
(iii) a negative impact on the return on investment for research and development expenditures, which can result in a competitive disadvantage to those enterprises which benefit from free-ride effects.

Moreover, counterfeit trade can

(iv) result in an increasing number of liability claims due to defective imitation products, and
(v) facilitate the emergence of future competitors since counterfeiting can help illicit actors to gather know-how that may enable them to become lawful enterprises in the future.

These implications explain the vivid interest in organizational and technical protection measures – especially since established security features have apparently not been able to prevent the increase in occurrences of counterfeiting. RFID technology has the potential to overcome the shortcomings of the established technologies and may become a powerful tool for product and brand protection. However, the wide range of affected products and industries, the large number of stake holders, and, last but not least, the considerable reengineering capabilities of many illicit actors require a thoughtful solution design.

In the following section, we introduce a non-formal attack model that helps to structure the requirement specification and the design phase. Thereafter, we outline and critically assess four principal solution concepts and show a possible migration path.

12.2   The use of RFID to avert counterfeit trade

The specification of auto-ID-based anti-counterfeiting technologies is strongly influenced by security-related requirements as well as by the design parameters which stem from an integration in the desired production and inspection settings; both aspects are discussed below.

The attack model

A critical design parameter of anti-counterfeiting technologies is the desired level of security. Attack models allow structuring of the requirement analysis. In cryptography, such models usually take the form of an "experiment," a program that intermediates communications with a fictional adversary, and a runtime environment containing the system components (often referred to as oracles) [3]. Security models have to accurately reflect real-world threats (i.e. the capabilities of illicit actors) as well as the actual system characteristics. With respect to RFID, appropriate models should address not only the top-layer protocols, but also the basic characteristics of RFID transponders down to the bit level. The latter may lead to less formal descriptions but is necessary in order to capture relevant threat scenarios like power analyses (and other side-channel attacks) and destructive reengineering tests. Therefore, the attack model outlined below consists of a non-formal description of the system characteristics and the capabilities of the illicit actors as well as the identification and evaluation of the – partly novel – attack scenarios.

System capabilities

Low-cost RFID transponders are limited with respect to their maximum gate count (as the chip size influences transponder cost), the available energy (due to the restrictions on transmitting power of readers, the size of the antenna, and the considerable distance between tag and reader devices which is often required), and the frequency spectrum. This ultimately results in limited computational power, confines the memory size and communication bandwidth, and hampers the integration of sophisticated pseudo-random-number generators or sensors against hardware attacks. Strong public key cryptography systems, for example, are still out of scope for low-cost RFID transponders.

Another important characteristic of RFID results from the radio connection between tag and reader. Connectivity is connectionless and communication is provided over an unreliable channel. This allows illicit actors to listen to the data exchange and, for example, retrieve existing identification numbers. Moreover, potential conflicts have to be considered when sharing the channel. Owing to the limited power of the readers and computational constraints among tags, a more powerful sender can easily jam legitimate readers [4]. An intentional violation of the tag-to-reader communication protocol, e.g. by continually transmitting messages in an attempt to generate collisions, can also disable a meaningful data exchange, which gives rise to several potential attacks.

Capabilities of illicit actors

The capabilities of illicit actors can by far exceed the computational power and hardware complexity of low-cost RFID transponders. Moreover, the unattended and distributed deployment of RFID transponders makes the devices highly susceptible to physical attacks. In fact, the access of the adversary to the system is a critical parameter of the attack model. Most cryptographic security analyses are based on the assumption that illicit actors are able to experiment extensively with the elements of the system [5] and thus are able to submit a large number of "oracle" queries to expose weaknesses of the design or to "guess" secret information. In this context, the limitations of RFID systems also restrict the capabilities of the attackers; illicit actors may have unlimited access only to selected transponders (e.g. after purchasing original articles with the security feature still in place), but limited access to arbitrary components. The latter is the case since attackers can read only those tags which are in close proximity to their reader devices, or listen to tag–reader communications that are taking place within eavesdropping range (see Juels [1] for a definition of various read ranges). However, in most supply-chain-related applications, the vast majority of transponders will remain hidden to other parties most of the time.

With respect to transponders that are in the possession of the attackers, a wide variety of tools is available. Potential steps include power analyses and the exact measurement of response times, the application of different clock speeds or the elimination of the air interface in order to increase the frequency of queries, and hardware attacks (e.g. opening the packaged IC) in the attempt to directly read out key registers on the circuit or to reverse-engineer the underlying algorithms. Therefore, when designing RFID-based anti-counterfeiting features, care must be taken that compromising accessible transponders does not affect the security of the remaining system. The protection should be based on secret keys that are different and non-related among the tags rather than on secret algorithms that a large number of transponders may have in common.

Attack scenarios

An attack model for low-cost RFID devices is provided by Juels [6], who mainly addresses threats to data security, authentication, and privacy. With respect to anti-counterfeiting features, however, the focus of potential attacks is shifted to an extended set of threats. Interviews with brand-protection experts conducted during this research revealed the relevance of the following issues: tag cloning, which is strongly related to tag authentication; obfuscation and deception; tag omission; removal–reapplication; and denial-of-service attacks. Each issue is addressed below.

1. *Tag cloning* refers to the duplication of security features such that they are likely to pass off as authentic during inspection. With respect to RFID, tag cloning may be defined as the replication of a transponder, with the duplicate being able to emulate the original tag's behavior. In a system with cloned entities, investigators (or reading devices), without taking the existence of duplicate features into account, would even falsely certify the authenticity of bogus components. In fact, large-scale tag-cloning attacks can severely compromise anti-counterfeiting solutions and therefore have to be addressed during system design.

2. *Obfuscation* connotes the use of misleading protection technologies. In practice, licit companies frequently change security features to prevent counterfeiters from copying or cloning their protection technology. While following this paradigm of "creating a moving target," the licit parties unintentionally complicate the inspection process. Especially third parties can be overwhelmed by the coexistence of different, mostly visual, security features. Consequently, counterfeit producers can often rely on the lack of knowledge (and the lack of time and motivation to acquire it) during inspection processes. A very common attack stems from the application of security mechanisms that are not related to the genuine product (e.g. the use of holograms instead of micro printings etc.). However, the need to change anti-counterfeiting primitives when they become ineffective as well as their user-friendliness given the limited resources during inspection translates into the requirement of a flexible security system with a static user interface.

    In anti-counterfeiting systems that rely on more than one component, threats may originate not only in bogus product security features but also in malicious back-end systems. When a barcode or an RFID transponder references a database containing track-and-trace information or advanced shipment notices, the authenticity of the relevant source has to be questioned.

3. *Tag omission*, i.e. the abdication of the security features by counterfeit producers even if the corresponding genuine articles are equipped with protective measures, relies on low inspection rates among many categories of goods. The phenomenon shows the need for large-scale and consequently low-cost inspection processes. Preferably, inspections can be automated even in loosely guided processes as implemented in many warehouses, at customs, or at retail stores.

4. *Removal–reapplication attacks* involve the application of genuine security features from (mostly discarded) genuine articles to counterfeits. They constitute a potential threat for tagging technologies in which security features are attached to an object (like holograms or RFID tags) rather than being an inherent part of

it (such as chemical markers). The consideration of this attack is of importance especially when protecting high-value goods such as aviation spare parts that are, when out of service, often still accessible to illicit actors. When relying on tagging technologies, a defense is to tightly couple the security feature to the object, e.g. by tamper-proofing its physical package or by providing a logical link between object and tag.

5. *A denial-of-service attack* may be defined as "any event that diminishes or eliminates a network's capacity to perform its expected function" [7]. Since established anti-counterfeiting technologies usually do not rely on network resources, this attack is new to the brand- and product-protection domain. However, when authentication processes involve entities in disparate locations, the access to these resources may be disturbed. With respect to RFID devices, attacks can cut off the connection between individual transponders and reading devices. When illicit actors target major distribution centers or customs, e.g. at harbors or airports, denial-of-service attacks may severely slow down inspection processes and thus interfere with the unobstructed flow of goods. Eliminating any possibility of such attacks is difficult, given the limited functionality of low-cost transponders. However, providing tools for detecting attacks and localizing the illicit device is not a major issue. In actual systems, the operator would have to localize and physically remove or deactivate the attack device.

Practical requirements

The attack model led to a set of security-related requirements. They include measures to avert a duplication of security features; the design of a stable, easy-to-use interface; the necessity for efficient inspection processes at low cost even in loosely guided processes; a tight coupling of the security feature to the object; and measures against denial-of-service attacks. In addition to this set, various – partly interrelated – conditions stem from the practical requirements on anti-counterfeiting solution which are not directly related to breaches of security. These include specifications from manufacturing (e.g. on the conditions a tag must endure during injection molding, integration into existing high-speed production systems etc.) and the often-required subordination of security techniques to product design (e.g. affecting the maximum size of transponders). The check-scenario-specific requirements also require special thought. With respect to RFID, bulk reading, read rates, read ranges, data standards etc. have to be considered.

12.3   Principal solution concepts based on RFID

RFID technology comes at various levels of complexity and can offer several functionalities that make it applicable as anti-counterfeiting measures.

Principle-application scenarios are plausibility checks based on unique serial numbers or track and trace, object-specific security, and secure product authentication.

Plausibility checks based on unique serial numbers

Marking objects with unique identifiers, i.e. on the item level rather than for individual project categories only, helps to monitor the flow of goods and thus to detect illicit trade activities. The basic operating principle of such a system is quite simple: the manufacturer generates a (random) number (ID), writes it to the data carrier and stores it in a database. When the product is checked, a reader device retrieves the ID and sends it to a service offered by the manufacturer, which looks up the number in the database; if the number is found to be valid, this can be interpreted as evidence for the authenticity of the product.

However, since IDs can be retrieved by everybody with a reader device and may be programmed into other RFID transponders, precautions have to be taken to hamper cloning attacks. In this context, the so-called tag ID can be used: the transponders produced to date contain an additional unique, read-only tag ID that is set during the manufacturing process (this is similar to the case of PC network cards, which also have a unique hardware address). Using the number pair serial number/tag ID leads to a considerably more secure solution without additional hardware expenses. In practical settings, two other important issues have to be considered: access management to prevent illicit actors or competitors from retrieving assigned numbers or information on the production volume; and address services to enable reading devices to localize the "identification database." Both issues have been addressed by EPCglobal working groups. [8]

Plausibility checks based on track and trace

In track-and-trace systems, information on an object's location and the corresponding time, possibly together with data on its owner and status, is recorded and stored for further processing. If such measurements are repeated over time, they allow plausibility checks of the product's history. Heuristics can be applied, e.g. as is done by credit card companies, which routinely block cards if they have a suspicious transaction history, in order to spot bogus products.

Track-and-trace systems rely on the ability to uniquely identify individual articles. In order to facilitate meaningful analyses, numerous data points have to be collected, which requires an efficient way to capture supply chain events. In this context, the unique serial number approach as outlined before can be seen as an enabling technology. Though the operating principles of track-and-trace systems may appear simple, actual implementation of the infrastructure is a severe challenge. From a technical perspective, especially access management in non-predetermined supply chains constitutes a major hurdle. However, even

Table 12.1. Product-specific security: system architecture

```
Product Validation Data := {
     Unique Tag ID,
     Unique Product Serial Number,
     Product Specific Data,
     Signature Method,
     Signature Value};
```

bigger barriers seem to be organizational issues concerning the ownership of the data, the distribution of system costs, and the lack of interest among some parties in providing their customer with a high degree of supply chain visibility. Nevertheless, track and trace is an important and promising technology, e.g. for the pharmaceutical industry, since the obstacles are to be seen alongside an effective way to avert counterfeiting, and enhanced production, inventory, and distribution control.

Object-specific security

Security solutions based on tagging technologies have a system-specific drawback: when checking an object, it is still the tag (e.g. a hologram or a simple RFID transponder) which is authenticated, not the object or document the tag is attached to. In other words, the link between tag and object is often not provided. In theory – and also in practice if the solution is not designed properly – a tag can be removed from an original article and attached to another object, thereby compromising the security system.

In contrast to most other tagging technologies, RFID can overcome this shortcoming. Even low-cost RFID tags with a certain amount of memory can store data that binds a tag to a given product, as a picture in a passport binds the document to its holder. An exemplary data set, termed product validation data, is given below, with the system architecture outlined (in Table. 12.1); the entries which are not self-explanatory are described briefly.

1. *Product-specific data.* This information resembles the picture or fingerprint in the passport analogy. The data has to be characteristic for an individual object, stable over time, and easily measurable during inspection. Which properties may be selected depends on the specific physical, chemical, electrical, etc. properties of a given object. Examples of characteristics are – either altogether or as a subset thereof – weight, electrical resistance, form factors, and a serial number printed on the product itself or its packaging. This data will typically be written on the tag by the product's vendor before product delivery, for example during packaging. It is also possible to store a reference to the data on the tag, such as an URI that specifies an entry in a remote database. This may help to save tag resources, but will make product validation dependent on the availability of network connectivity.

2. *Signature value*. The product vendor computes the signature value by applying a cryptographic hash function h with a public key encryption method SPr, such that

$$\text{Sig\_Value} = \text{SPr}(h(\text{Unique Product Identifier, Unique Tag Identifier,}$$
$$\text{Signature Method, Validation Key})).$$

Here, SPr indicates the usage of the vendor's private key (a.k.a. signing key) when computing the signature value. Note that the private key must be known exclusively to the entity (e.g. product vendor). During product validation, the corresponding public key called the validation key will be used to check the validity of the signature value.

3. *Validation key*. The validation key is a public key used to compute the signature value. It may be retrieved from a trusted online source, which can be localized over an address lookup service as outlined before.

An advantage of this approach is that low-cost tags with approximately 32–64 bytes of memory can be used. It does not rely on cryptographic functions on the tag, which would require more-expensive transponders. The approach can also be combined with plausibility checks based on track and trace or secure tag authentication principles to avert cloning attacks.

Secure authentication

Strong protection against cloning attacks requires secure authentication on the side of the tag. However, establishing efficient means of authentication in RFID infrastructures constitutes a major challenge. Since IDs are usually not read protected, an advanced attacker may be able to obtain an identifier from a tag and program it into another transponder, or emulate the tag using some other wireless device. If produced on a large scale by illicit actors, duplicate devices render track and trace or anti-counterfeiting solutions ineffective.

Challenge–response protocols can avert tag cloning since they allow comparison of secret keys at disparate locations without transferring them over a possibly insecure channel. Critics of this approach frequently mention the increasing tag costs which may result from the integration of the required cryptographic unit into RFID transponders. However, Feldhofer *et al.* [9] showed an implementation of a 128-bit version of the Rijndael cipher [10] using fewer than 4,000 gates, which, given an approximate gate count of current EPC GenII tags of 20,000, would lead to only a small increase in tag cost. However, there are other obstacles that are often overlooked: limitations on communication bandwidth and power supply. Other than just broadcasting an ID, authentication mechanisms additionally require the transmission of a challenge and a response, including steps to select and address individual tags. Moreover, the computationally intense operations need more energy over a longer period of time. Both significantly reduce read ranges and bulk-reading capabilities. However, for authentication of small quantities of high-value products, this does not constitute a severe drawback.

12.4   Migration paths and application scenarios

As can be inferred from the previous section, the level of security has a strong impact not only on the design but also on the fixed and variable costs of the solution. The desired level can be determined

(i) by the risk or cost resulting from a compromised system, and
(ii) by the lifetime of the object which is to be protected.

Risk and cost can be classified in terms of the potential health and safety hazards for consumers or the incremental financial losses of licit manufacturers and brand owners. With respect to RFID systems, the respective design space ranges from low-cost ID tags to transponders with advanced authentication capabilities.

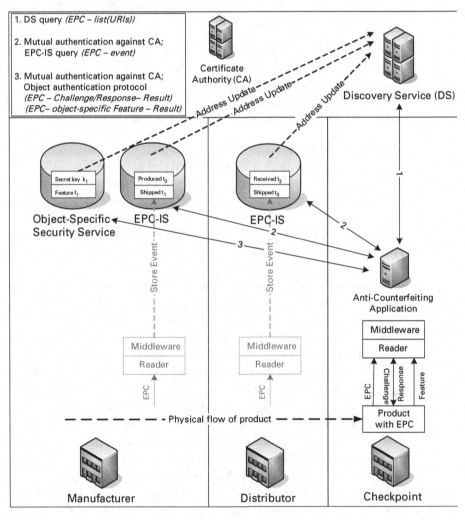

Fig. 12.1   Infrastructure for different security techniques.

In fact, several different security solutions will coexist. This is desirable also because anti-counterfeiting technologies often constitute a barrier to illicit actors for only a limited period of time. Consequently, it is desirable to have the opportunity to change the underlying security primitive at low cost, i.e. without the need to alter the technical infrastructure or to require the user to get accustomed to new checking procedures. In this context, the use of such features to support customs constitutes an insightful application scenario. There, numerous articles with various security requirements are to be inspected at high speed. According to interviews with customs officials, one hand-held device has to be sufficient for inspecting all goods; relying on a "battery" of reading devices would be impractical. This gives rise to the logical next step: the design of one standardized infrastructure that facilitates the different RFID-based security techniques (cf. Fig. 12.1). The EPC network can provide the foundation for future work.

12.5   Conclusion

In this chapter, we outlined the requirements and principal techniques to avert counterfeit trade using RFID. We introduced an attack model to categorize and describe potential threats and thereby help to structure the design process. The relevance of obfuscation and tag omission turned the attention to issues beyond typical IT security since it leads one to stress the importance of systems that can be used with low effort. Since RFID can facilitate large-scale, automated product checks at low cost, it is well suited for enhancing product security in practical settings. The existence of a migration path towards more secure approaches without the need to change the underlying inspection routines contributes to the sustainability of the solution. Future research is nevertheless required, especially in the fields of access management and tag authentication, and, last but not least, to integrate the system into one standardized network infrastructure.

12.6   References

[1] **TAXUD European Taxation and Customs Union**, *Breakdown of the Number of Cases Registered and the Number of Articles Seized by Product Type* (European Union, 2004) (http://ec.europa.eu/taxation_customs/resources/documents/customs/customs_controls/ counterfeit_piracy/statistics/counterf_comm_2004_en.pdf).

[2] **International Chamber of Commerce (ICC)**, *Current and Emerging Intellectual Property Issues for Business – A Roadmap for Business and Policy Makers*, revision 6 (ICC, Paris, 2005), p. 2 (www.insme.info/documenti/Roadmap-2005-FINAL.pdf).

[3] **Juels, A.**, "RFID Security and Privacy: A Research Survey," *IEEE Journal on Selected Areas in Communications*, 24(2):381–394 (2006).

[4] Walters, J. P., Liang, Z., Shi, W., and Chaudhary, V., "Wireless Sensor Network Security: A Survey," in *Security in Distributed, Grid, and Pervasive Computing*, ed. Y. Xiao (CRC Press, Boca Raton, FL, 2006).

[5] Bellare, M., Desai, A., Pointcheval, D., and Rogaway, P., "Relations among Notions of Security for Public-key Encryption Schemes," in *Proceedings of the 18th Annual International Cryptology Conference (CRYPTO '98)*, ed. H. Krawczyk, pp. 26–45 (Springer-Verlag, Berlin, 1998).

[6] Juels, A., *Minimalist Cryptography for Low-cost RFID Tags* (RSA Laboratories, Bedford, MA, 2004) (www.rsasecurity.com/rsalabs/staff/bios/ajuels/publications/minimalist/Minimalist.pdf).

[7] Wood, A. D., and Stankovic, J. A., "Denial of Service in Sensor Networks," *Computer*, 35(10):54–62 (2002).

[8] Traub, K., Allgair, G., Barthel, H. *et al.*, *The EPCglobal Architecture Framework* (EPC-global, Princeton, NJ, 2005) (www.epcglobalinc.org/standards/Final-epcglobal-arch-20050701.pdf).

[9] Feldhofer, M., Dominikus, S., and Wolkerstorfer, J., "Strong Authentication for RFID Systems Using the AES Algorithm," in *Proceedings of the Workshop on Cryptographic Hardware and Embedded Systems (CHES '04)*, pp. 357–370 (Springer-Verlag, Berlin, 2004).

[10] Daemen, J., and Rijmen, V., *The Design of Rijndael: AES – The Advanced Encryption Standard* (Springer-Verlag, Berlin, Berlin, 2002).

13 Closing product information loops with product-embedded information devices: RFID technology and applications, models and metrics

Dimitris Kiritsis, Hong-Bae Jun, and Paul Xirouchakis

As companies seek to control their brands and products across ever-expanding global supply chains, further visibility can be extended beyond the point of sale to the points of use across the product lifecycle. *Closed-loop product lifecycle management (closed-loop PLM)* focuses on tracking and managing the information of the whole product lifecycle, with possible feedback of information to product lifecycle phases. Implementing the PLM system requires a high level of coordination and integration of the product-embedded information devices (PEIDs) that track the products. To fulfill this need, the concept of *closed-loop PLM*, its system architecture, modeling framework, and metrics have been addressed.

13.1 Introduction: closing the product information loop

From a product lifecycle management perspective, a product's lifecycle can be separated into the following main phases: beginning of life (BOL), including design and production: middle of life (MOL), including use, service, and maintenance: and end of life (EOL), characterized by various scenarios such as reuse of the product with refurbishing, reuse of components with disassembly and refurbishing, material reclamation without disassembly, material reclamation with disassembly, and, finally, disposal with or without incineration.

We may say that between the first two phases, i.e. design and production, the information flow is quite complete and supported by intelligent systems such as CAD/CAM, product data management (PDM), and knowledge management systems that are effectively and efficiently used by the industry. However, the information flow becomes less and less complete from the MOL phase to the final EOL

RFID Technology and Applications, eds. Stephen B. Miles, Sanjay E. Sharma, and John R. Williams. Published by Cambridge University Press. © Cambridge University Press 2008.

scenario. It is acknowledged that, in most cases, the information flow breaks down after the delivery of the product to the customer. The fact that the information flow is in most cases interrupted shortly after product sale restricts the feedback of data, information, and knowledge from service, maintenance, and recycling experts back to designers and producers. That is why one of the main objectives of PROMISE [1] is to develop appropriate technology and associated information models in order to close the information gap.

In the closed information loops, several research issues can be newly highlighted. For example, "Design for X" at BOL can take into account during the design phase not only the final product but also other non-functional aspects related to product lifecycle. In "Design for X," "X" stands for "manufacture," "assembly," "disassembly," "quality," "safety," "recycling," "environment," "variety," "reliability," "end of life," "maintenance," etc., depending on the aspect taken into account during the design. These aspects overlap to a greater or lesser degree and consequently are not necessarily independent. For example, with regard to maintainability, relevant attributes may be considered during the early phase of the design process, in the closed information loops environment, simultaneously with other lifecycle attributes like reliability, manufacturability, assemblability, safety, etc. Furthermore, in the closed information loops environment, the maintenance data collected during product MOL can be used for various purposes such as maintenance management, design for maintenance or the implementation of an e-maintenance system for a remote maintenance service. The product usage data during MOL can be collected either automatically using sensors or similar devices, or by the experts involved in activities related to product MOL such as maintenance. Finally, in the closed information loops environment, the PEID can greatly increase the effectiveness of EOL management. For example, material recycling can be significantly improved because recyclers and re-users can obtain accurate information about "value parts and materials" arriving via EOL routes: what materials they contain, who manufactured them, and other knowledge that facilitates material re-use [2].

These concepts can be implemented in a *closed-loop PLM* system. In general, to build up an enterprise-level system like the PLM system, it is important to address the requirements of total system integration during an early stage of development in order to reduce the risks of cost and schedule overruns. This requires a system architecture. The system architecture can be defined as the organizational structure of a system which provides standards and techniques for building complex functions of a system. It is being recognized as a critical process in developing complex systems because it simplifies the management of complex environments, provides coherence and consistency, and ensures that user requirements are met. Moreover, it helps us to understand the boundary and system requirements clearly. Hence, to coordinate and efficiently manage complex PLM-implementing activities, it is necessary first and foremost to develop a generic system architecture for PLM in a standard and flexible way so that it can be adapted to various application domains. One should address how the PLM system and its context will be represented in viewpoints of system components and their relations. This architecture is very important because it is a basic sketch for developing a *closed-loop PLM* system. It can provide people with a common language in which to seek consensus over how a PLM system should be built.

In this chapter, we describe the concept of *closed-loop PLM* and its system architecture developed in the framework of the EU FP6 507100 and IMS 01008 project PROMISE [1] [2] [3]. It contains the definition of necessary components for the *closed-loop PLM* and how to integrate and coordinate its components with respect to hardware and software. In addition, we address a framework for modeling product usage data with UML and RDF.

13.2   The concept of closed-loop PLM

Recently, with emerging technologies such as wireless sensors, wireless telecommunication, and product identification technologies, product lifecycle management (PLM) has been in the spotlight. PLM is a new strategic approach to managing the product-related information efficiently over the whole product lifecycle. Its concept appeared in the late 1990s, moving beyond the engineering aspects of a product and providing a shared platform for creation, organization, and dissemination of product-related knowledge across the extended enterprise [4]. PLM is defined as follows.

A strategic business approach that applies a consistent set of business solutions in support of the collaborative creation, management, dissemination, and use of product definition information across the extended enterprise from concept to end of life integrating people, processes, business systems, and information [5].

PLM facilitates the innovation of enterprise operations by integrating people, processes, business systems, and information throughout the product lifecycle and across the extended enterprise. It aims to derive the advantages of horizontally connecting functional silos in the organization, enhancing information sharing, facilitating efficient change management, allowing use of past knowledge, and so on [6].

In particular, thanks to recent product identification technologies such as radio frequency identification (RFID) [7] and AUTO-ID [8], the whole product lifecycle can now be made visible and controllable. This allows all actors involved in the whole product lifecycle to access, manage, and control product-related information, especially the information after delivery of the product to the customer and up to its final destiny, without temporal and spatial constraints. During the whole product lifecycle, we can have visibility of not only forward but also backward information flow. For example, BOL information can be used to streamline MOL and EOL operations. Furthermore, MOL and EOL information can also go back to the designer and producer for improvement of BOL decisions. This indicates that the information flow is horizontally closed over the whole product lifecycle. In addition,

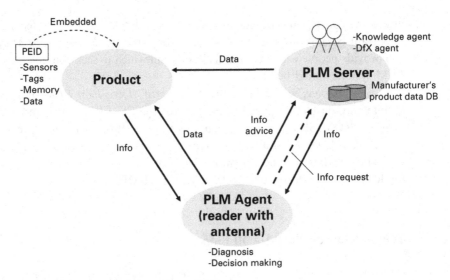

Fig. 13.1   The overall concept of closed-loop PLM.

information flow is vertically closed. This means that, with data gathered by sensors, we can analyze product-related information and take some decisions on the behavior of products, which will affect data gathering again. We call this concept and relevant systems closed-loop PLM in this study. The concept of closed-loop PLM (illustrated in Fig. 13.1 above) can be defined as follows.

A strategic business approach for the effective management of product lifecycle activities by using product data/information/knowledge accumulated in the closed loops of product lifecycle with the support of a product-embedded information device (PEID) and product data and knowledge management (PDKM) system.

The objective is to streamline product lifecycle operations over the whole product lifecycle, on the basis of seamless product information flow, through a local wireless network of PEIDs and through remote internet connection to knowledge repositories in PDKM.

To implement the concept of closed-loop PLM, the following are necessary conditions.

- Every product has a PEID to manage its lifecycle data. If necessary, sensors can be built into products and linked to the PEID for gathering its status data.
- Each lifecycle actor has accesses to PEIDs with its reader or to a remote PLM system for getting necessary information.
- Closed-loop PLM should have decision-support systems and PDKM systems for providing lifecycle actors with suitable advice at any time.
- In closed-loop PLM, information flow is horizontally closed, which means that information flow covers all product lifecycle phases: BOL, MOL, and EOL.
- Designers at BOL will be able to exploit expertise and know-how of the other actors in MOL and EOL, such as the modes of use, conditions of retirement,

and disposal of their products, and thus will be able to improve product designs.

- Producers at BOL will be provided in a real-time way with not only operational data from the shop floor but also usage status of a product until the disposal phase.
- Service and maintenance experts at MOL will be assisted in their work by having not only product design information at BOL but also an up-to-date report about the status of the product during the product usage period at MOL.
- Recyclers and re-users at EOL will be able to obtain accurate information about "value materials" arriving through EOL routes by analysis of modes of use and conditions of a product at MOL based on product specification at BOL.

Moreover, the information flow is vertically closed, which means that product lifecycle information is controlled in the vertical loops of gathering, analyzing, using, and adjusting data gathering again.

- A PEID gathers product-related data under specific conditions or periodically or in a real-time way over the whole product lifecycle.
- A PEID sends gathered data to a database under specific conditions or periodically or in a real-time way.
- From the data gathered, information and knowledge are generated and stored at a knowledge repository in a PDKM system. They are used for the making of decisions by lifecycle actors.
- After analysis and decision making, if there is any need to update product information or adjust data gathering, the PLM server sends PLM agents or directly sends messages to the PEID.

13.3   The state of the art

There has been some work dealing with modeling issues relevant to PLM. For example, CIMdata [5] addressed a high-level PLM definition, describing its core components and clarifying what is and what is not included in a PLM business approach. CIMdata mentioned three core concepts of PLM: (1) universal, secure, managed access and use of product definition information; (2) maintaining the integrity of product definition and related information throughout the life of the product or plant; and (3) managing and maintaining business processes used to create, manage, disseminate, share, and use the information. Ming and Lu [9] proposed a new business model in a virtual enterprise in order to tackle issues of product development within the scope of PLM. They proposed a framework of product lifecycle process management for collaborative product services. The framework consists of an industry-specific product lifecycle process template, product lifecycle process application, abstract process lifecycle management, supporting process technology, supporting standards, and enabling infrastructure. Moreover, Gsottberger *et al.* [10] developed a hardware and software architecture (called *Sindrion*) to integrate simple, low-power, and low-cost sensors and actuators into universal plug and play (uPnP) environments for ubiquitous computing. It defines interfaces for a wireless point-to-point connection between so-called *Sindrion transceivers* and certain computing terminals. The Sindrion architecture focuses on maintenance and control applications for white goods. In addition, Georgiev and Ovtcharova [11] recently proposed an architecture to support easy prototyping of PLM *n*-tier systems based on web services. In their work, they have focused on creating reliable prototypes of control web services for a PLM system.

13.4   System architecture

In this section, we describe the system architecture of closed-loop PLM to clarify its concept by defining necessary components and their relations. It has multiple layers as shown in Fig. 13.2 [12], which are mainly classified into business process, software, and hardware. The PEID can be regarded as an important hardware component for facilitating the closed-loop PLM concept. Furthermore, software related to applications and middleware layers, and their interfaces, plays important roles in closed-loop PLM. Hence, we introduce PEID and software architectures in the next sections.

Hardware architecture: PEID

PEID stands for "product-embedded information device." It is defined as a device embedded in (or attached to) a product, which contains its information

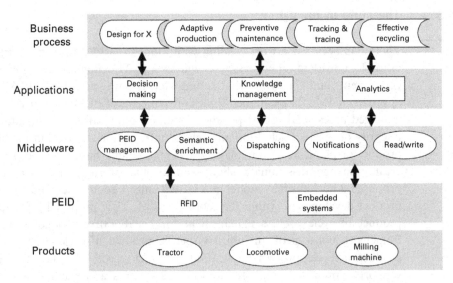

Fig. 13.2   An overview of closed-loop PLM system architecture (the PROMISE case).

(e.g. product identification) and is able to provide the information whenever requested by externals during the product lifecycle. There are various kinds of information devices built into products to gather and manage product information. We group them into two types: RFID tags and on-board computers. In this study, we use the term "PEID" to cover all things. The term "product-embedded" means that product lifecycle information can be tracked and traced in a real-time way over the whole product lifecycle by attaching a PEID to a product itself. For this, a PEID should possess a unique identity and its work should not be dependent on the availability of power to it. This requires product identification and power management functions in a PEID. The term "information device" indicates that a PEID should have data processing, data storing, and sensor reading functions. These functions enable a PEID to gather data from several sensors; to retain or store them; and, if necessary, to analyze them or support the making of decisions. In addition, it should have a communication function with external environments for exchanging data. For this, a PEID should have a processing unit, communication unit, sensor reader, data processor, and memory.

Software architecture

PLM has emerged as an enterprise solution. It implies that all software tools/systems/databases used by various departments and suppliers throughout the whole product lifecycle have to be integrated so that the information contained in these systems can be shared promptly and correctly between people and applications [6]. Hence, it is important to understand how application software in PLM fits with others in order to manage product information and operations [5]. For this, a software architecture is required. Software architecture is the high-level structure of a software system concerned with how to design software components and make them work together. In this study, we focus on structural views of software architecture in terms of components and their relationships.

Figure 13.3 shows the software architecture for closed-loop PLM. It has a vertical viewpoint in the sense that its structure represents a hierarchy of software of closed-loop PLM from gathering raw data to business applications. For example, embedded software (called firmware) built into a PEID has the role of controlling and managing PEID data. Database software is also required to store sensor data and manage it efficiently. Middleware can be considered in general as intermediate software between applications. It plays a role as the glue between software layers. It is used to support complex, distributed applications, for example, applications involving interactions between RFID tags and business information systems to communicate, coordinate, and manage gathered data.

PDKM manages information and knowledge generated during the product lifecycle. It is generally linked with decision-support systems and data-transformation software. PDKM is a process and the associated technology needed to acquire, store, share, and secure understandings, insights, and core distinctions. PDKM should link

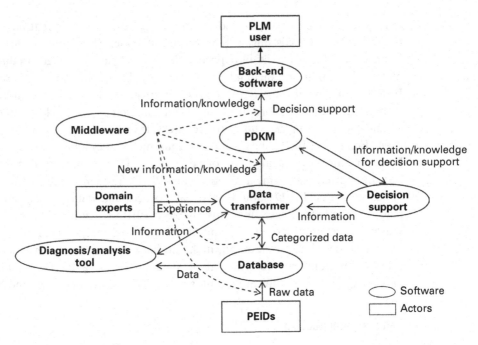

Fig. 13.3   Software architecture.

not only product design and development such as CAD/CAM but also other back-end software (legacy systems), e.g. enterprise resource planning (ERP), supply chain management (SCM), and customer relationship management (CRM), to achieve interoperability of all activities that affect a product and its lifecycle. PDKM is very important in closed-loop PLM because it generates and manages core properties of the product lifecycle in order to acquire competence. Back-end software can be defined as the part of a software system that processes the input from the front-end system dealing with the user.

13.5   A business case of PROMISE on ELV recovery

End-of-life vehicle (ELV) recovery is on the agenda of the automotive industry today. In the past the majority of used products were landfilled or incinerated, with considerable damage to the environment. It is estimated that 10% of all hazardous waste generated in Europe originated from ELVs. Landfill costs have recently increased steadily and are expected to continue to rise. Moreover, many products can no longer be landfilled because of environmental regulations. In the EU, new laws dictate that the producers of certain products such as vehicles must bear responsibility for their final disposal. These developments have given rise to a new material flow from the end user back to the producers.

In this context the following challenges are identified.

- How can the dismantler of a car know which parts are worth removing for re-use and refurbishment, which need special treatment to meet environmental regulations, and which should go to the shredder?
- How can the manufacturer prove it is meeting the ELV directive targets of the European Union (2000/53/EC)?

PROMISE may help to provide solutions to such challenges in the following way: after a vehicle has reached its end of life and been delivered to its take-back point, its basic information – such as the type of vehicle, its ID, and the assembly date – are taken from a top-level PEID. Mission profile information and statistics about the use of the vehicle and its components, e.g. kilometers driven and environmental conditions such as humidity, external temperature, temperature in the engine area, etc. may be available in the PEID. With this information and the maintenance history, parts and components worth re-using or remanufacturing are identified and removed for further use with or without refurbishing. Other components such as glass, bumpers, foam, etc. are removed for recycling. This helps the dismantler in minimizing disassembly time and effort, and increasing earnings from optimal use of parts with a remaining life and value at the EOL of the vehicle.

Since PROMISE is "closing the product information loops," the benefits of using it are not limited to the above. ELV information is fed back to other stakeholders in the automotive product lifecycle chain: first of all, engineering and manufacturing are provided with important information about the vehicle at the end of its life, helping to identify, for example, problems with the design of some components or problems with disassembly operations; on the other hand, marketing may get precise information about compliance with the ELV Directive. From the enterprise performance management viewpoint, the implementation of the PROMISE concept (closed-loop PLM) will give many benefits. With the information visibility attained by the use of PEIDs, enterprise performance management can be activated. In closed-loop PLM, measuring performance indicators (e.g. throughput rate and lead time) for each lifecycle operation can be done automatically and accurately. This facilitates smart decision making on other operational and managerial issues throughout the whole product lifecycle.

More information on the PROMISE business cases may be found in the white paper [13] on PROMISE and in descriptions of case studies [14].

13.6   Product usage data modeling with UML and RDF

In closed-loop PLM, enormous amounts of product usage data are gathered and accumulated. Each item of PEID data has just a simple meaning. It cannot tell us about product status or expected problems with the product. Hence, there is a challenging issue: that of how to manage and design product usage data in a ubiquitous environment.

Product usage data in closed-loop PLM has the following characteristics.

- Product usage data is created by time-based events. Some items of product usage data are automatically recorded by tagging PEIDs (simply RFID tags) of products, while some are manually recorded by an engineer's operations. Irrespective of the means of recording, all data is generated on the basis of specific time events.
- A huge amount of product usage data is usually generated since there are lots of data-gathering probes in the ubiquitous environment.
- Each item of data can have a unique identification. A PEID can be attached to each lifecycle object such as product, process, and resource. The PEID can have a unique identification scheme following the "de facto" standard of a product identification scheme (e.g. EPC of auto-ID).

To capture the characteristics of product usage data, the following are required.

- It is necessary to summarize and consolidate the huge amount of PEID event data.
- It is necessary to help engineers to design and manage product usage data in an easy and simple manner.
- The information model should be flexible so that it is easy to add to and extend it since the product usage data will change over time.
- The information model should be shared in an easy way among involved partners.
- The information model should help one to find necessary data and information in an efficient manner.

For this, it is necessary to develop a method for modeling product usage data with the goal of efficiently providing necessary information to customers and maintenance engineers. To this end, we can apply unified modeling language (UML) diagrams and a resource description framework (RDF).

The UML is a general-purpose visual modeling language for developing a software system. Because of its usefulness, it is now used not only in software development but also in business process engineering. On the other hand, RDF is a standard language recommended by W3C for representing meta data about not only web resources but also other resources that cannot be directly retrieved on the web. RDF is a suitable language for describing product usage data because it can describe the relations among all objects and their properties during the product usage period with a simple representation method.

Figure 13.4(a) shows the overall framework for modeling product usage data with UML and RDF. The proposed framework consists of three layers: UML modeling, extracting and transforming, and RDF model layers. The framework enables non-experts to design and manage product usage data in a stepwise manner.

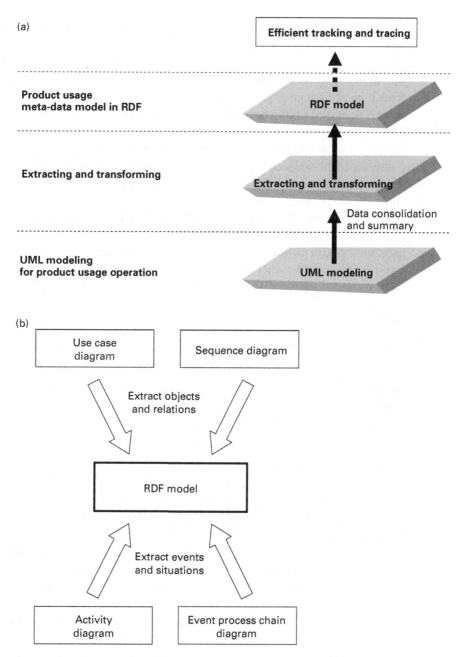

Efficient tracking and tracing

Product usage
meta-data model in RDF

RDF model

Extracting and transforming

Extracting and transforming

Data consolidation
and summary

UML modeling
for product usage operation

UML modeling

(b)

Use case
diagram

Sequence diagram

Extract objects
and relations

RDF model

Extract events
and situations

Activity
diagram

Event process chain
diagram

Fig. 13.4 The product data usage model.

Figure 13.4(b) shows the UML modeling layer, where we apply four UML diagrams to build up the RDF model for product usage data. To extract objects and relations from the product usage domain, a use case diagram and a sequence diagram can be applied. They can be a basis to design the RDF schema model for product usage data. On the other hand, an activity diagram and an event process chain (EPC) diagram can be used to extract those events which are prerequisites for summarizing and consolidating a huge amount of PEID data. The amount of data generated during the product usage period is enormous. Each item of data gives just simple information, which is not meaningful as long as each item is not related to others. This means that we should look at series of data for finding meaningful information. This requires an elaborate method to analyze, consolidate, and summarize enormous amounts of data, and infer new events for identifying and providing meaningful data and information. For this, we need to enumerate which events can happen by drawing activity diagrams and EPC diagrams.

Throughout the extracting and transforming layer, enormous amounts of PEID data should be updated into an RDF model. Through activity diagrams and EPC diagrams, we can identify conditions that trigger specific events. On that basis, we can infer important events among lots of PEID data. Events can be updated by RDF queries. For example, the following is one case using SeRQL.

```
SELECT *
FROM {product} is_recorded_at {t1};
has_properties {β},
{Actor} execute_at {t2};
has_properties {a}
WHERE (t1 = t2) AND (the predefined rules are satisfied)
CONSTRUCT
{Process} is_executed_at {t1};
has_property {γ}
```

Finding events through analyzing PEID tagging data can provide meaningful information. However, to facilitate application of the information, the information should be stored in certain forms somewhere. Generated events can be updated into an RDF model for product usage data, which is the role of the RDF modeling layer.

The use of UML diagrams is helpful in the following points. First, UML's graphical notation makes it easy for technical and non-technical people to share common concepts of a product usage environment. Second, UML diagrams can represent the relations between objects involved in product usage operations, which facilitates building up an RDF model. We can extract objects and relations from a use case diagram and a sequence diagram. The RDF schema model can be made on the basis of extracted objects and relations. Finally, we can extract events from an activity diagram and an EPC diagram. This information can be used to summarize and consolidate enormous quantities of usage data. The summarized information can be contained in the RDF model.

On the other hand, RDF modeling has the following benefits in product usage data modeling. The RDF model is a practical way to describe product usage data. In the ubiquitous environment, product usage data related to products, processes, resources, and agents is usually generated by accessing PEIDs periodically or randomly by triggering of some events, which are recorded together with the time. The RDF model deals with these main objects and their relations. Furthermore, not only physical lifecycle objects having PEIDs (e.g. products, resources) but also conceptual objects such as processes, agents, and time stamps can have unique identifications, which are compatible with the RDF representation scheme. In RDF, one incorporates the link structure between uniform resource identifiers (URIs). Hence, we can support the consistency of data semantics with unique IDs of resources. Furthermore, the model is flexible with respect to extension or modification. It is simple to import or export RDF statements (subject, predicate, and object triples) from any kind of database. In closed-loop PLM, product usage data is generated from various places in a distributed manner. Product usage data also changes over time. Sometimes, we do not know what we might want to combine it with next. In those cases, using RDF is useful. In addition, it is possible to efficiently look up product lifecycle data by tracing link structures between URIs of RDF statements if we can develop RDF query applications.

13.7   Conclusion

In this chapter, we have proposed the concept of closed-loop PLM and its system architecture, in which information flow is horizontally and vertically closed. For this, we have described the system architecture from two viewpoints: hardware and software. With the system architecture, we can get an opportunity to create leverage and synergies, and avoid duplication and inconsistency when we build up a complex information system across the enterprise. In addition, we have addressed a framework for modeling product usage data with UML and RDF. Although our work might not provide the exhaustive result for the closed-loop PLM model, we think that our work has laid the cornerstone for studying closed-loop PLM.

13.8   Acknowledgments

The content of this chapter is based on the PROMISE project (EU FP6 507100 and IMS 01008) that is currently under development (www.promise-plm.com). We wish to express our deep gratitude to all PROMISE partners.

13.9   References

[1] **Kiritsis, D.**, **Bufardi, A.**, and **Xirouchakis, P.**, "Research Issues on Product Lifecycle Management and Information Tracking Using Smart Embedded Systems," *Advanced Engineering Informatics*, 17:189–202 (2003).

[2] **PROMISE**, *PROMISE – Integrated Project: Annex I – Description of Work* (2004) (http://www.promise-plm.com).

[3] **Kiritsis, D.**, and **Rolstadås, A.**, "PROMISE – A Closed-loop Product Lifecycle Management Approach," in *Proceedings of IFIP 5.7 Advances in Production Management Systems: Modeling and Implementing the Integrated Enterprise* (2005).

[4] **Ameri, F.**, and **Dutta, D.**, *Product Lifecycle Management: Needs, Concepts and Components* (Product Lifecycle Management Development Consortium, 2004).

[5] **CIMdata Inc.**, *Product Lifecycle Management – Empowering the Future of Business* (CIMdata Inc., 2002).

[6] **Macchi, M.**, **Garetti, M.**, and **Terzi, S.**, "Using the PLM Approach in the Implementation of Globally Scaled Manufacturing," in *Proceedings of International IMS Forum 2004: Global Challenges in Manufacturing* (2004).

[7] **Schneider, M.**, *Radio Frequency Identification (RFID) Technology and Its Application in the Commercial Construction Industry* (University of Kentucky, 2003).

[8] **Parlikad, A. K.**, **McFarlane, D.**, **Fleisch, E.**, and **Gross, S.**, *The Role of Product Identity in End-of-life Decision Making* (Auto-ID Center, Institute of Manufacturing, Cambridge, 2003).

[9] **Ming, X. G.**, and **Lu, W. F.**, "A Framework of Implementation of Collaborative Product Service in Virtual Enterprise," in *Proceedings of Innovation in Manufacturing Systems and Technology (IMST)* (2003).

[10] **Gsottberger, Y.**, **Shi, X.**, **Stromberg, G.**, **Sturm, T. F.**, and **Weber, W.**, "Embedding Low-Cost Wireless Sensors into Universal Plug and Play Environments," in *Proceedings of 1st European Workshop on Wireless Sensor Networks (EWSN 04)* (2004), pp. 291–306.

[11] **Georgiev, I.**, and **Ovtcharova, J.**, "Modeling Web-services for PLM *N*-Tier Architecture," in *Proceedings of International Conference on Product Lifecycle Management* (2005), pp. 199–209.

[12] **Hackenbroich, G.**, and **Nochta, Z.**, "A Process Oriented Software Architecture for Product Lifecycle Management," in *Proceedings of 18th International Conference on Production Research* (2005).

[13] **Stark, J.**, *The Promise of Increasing Business Value with PLM and Smart Products* (2006) (http://www.promise-plm.com).

[14] **PROMISE**, *PROMISE Case Studies* (2006) (http://www.promise-plm.com).

14 Moving from RFID to autonomous cooperating logistic processes

Bernd Scholz-Reiter, Dieter Uckelmann, Christian Gorldt, Uwe Hinrichs, and Jan Topi Tervo

14.1 Introduction to autonomous cooperating logistic processes and handling systems

During the last few decades the structural and dynamic complexity in logistics and production has increased steadily [1]. Many causes for higher structural complexity can be found, for instance, in the integration of multiple companies in production and logistic networks. This effect is furthermore amplified by a growing internal and dynamic complexity caused, for example, by an increasing number of product variants. Likewise, dynamic customer behavior intensifies this situation [2]. All these effects combined lead to higher information requirements.

For efficient planning and control a broad and reliable basis of information is needed [3]. However, the underlying algorithms will soon face the end of computation capacity due to the large amount of information that has to be taken into account. It is foreseeable that in the future centralized planning and control methods will not be able to process all the information delivered. A solution to this dilemma is the decentralized storage of necessary information on the logistic object itself as well as the capability of local decision-making. In order to achieve this goal, logistic objects themselves have to become intelligent.

The emergence of these intelligent objects is the foundation for autonomous cooperating logistic processes [4]. The main idea of this concept is to develop decentralized and heterarchical planning and control methods as opposed to existing centralized and hierarchical planning and control approaches. It requires that interacting elements in non-predictable systems possess the ability and the possibility to render decisions independently. Autonomous control thus aims at an increased robustness as well as positive emergence of the complete system through distributed and flexible coping with dynamics and complexity [5]. The main point of this concept is to augment all logistic objects with a certain degree of intelligence in order to store and process relevant data, communicate and interact with

RFID Technology and Applications, eds. Stephen B. Miles, Sanjay E. Sharma, and John R. Williams. Published by Cambridge University Press. © Cambridge University Press 2008.

the other logistic objects, and get information about the surrounding environment and interact with it.

Data storage and processing is needed for decision-making locally. Data can be categorized into static and dynamic, depending on the acquisition. Static data can be characterized as relevant data that does not change over time, such as an ID or material properties, or changes only over long intervals, such as lot or production line, and stays constant within the supply chain or the product lifecycle of a single product.[1] Dynamic data can be acquired via sensors and changes over time in terms of temperature and positioning information.

Communication is also a very important element of the decentralized decision-making process. All logistic objects, products as well as machines and handling systems, have to share relevant information with each other in order to negotiate and find the best possible local and global solution. Sensing abilities are needed in order to gain information on the surrounding environment and include this knowledge in the decision-making process. For instance, critical parameters such as temperature and humidity are needed in order to calculate remaining shelf-life for certain products. The temporal evolution of the critical parameter can be dynamically stored in the transponder memory and then retrieved when needed for decision-making.

The autonomy of logistic objects, such as goods and returnable transport items, is enabled by new information and communication technologies that provide sensing, data storage, and data processing abilities as well as the ability to communicate with logistic objects. Radio frequency (RF)-based technologies, wireless communication networks, and software agents are the main drivers to permanently identify and locate these intelligent entities within the logistic system and enable communication with and among them [6].

These technologies are fundamental for intelligent logistic objects to act autonomously, but autonomous control does not just include intelligent logistic objects. Even though decision-making is decentralized, logistic objects still lack the capability to execute decisions. Moreover, automated handling systems such as automated sorters, which are nowadays in service to route logistic objects (e.g. parcels) to their correct destinations, have to be augmented by functions enabling them to communicate with the intelligent objects to execute their decisions. In this way, fully autonomously controlled processes can be achieved on every logistic level from single goods via pallets and containers to vehicles always being able to request special handling. The coalescence of material and information flow (information processing and decision execution) will enable every item or product to manage and control its logistic process autonomously.

This chapter will discuss the convergence of RF-based technologies and the concept of autonomous cooperating logistic processes. It will extend this approach by integrating automated handling systems as a means to accomplish the intelligent and autonomous handling of logistic objects. The key benefits of RFID and other RF-based technologies which enable logistic objects to become

---

[1] Some researchers define this kind of data as "semi-dynamic" [2].

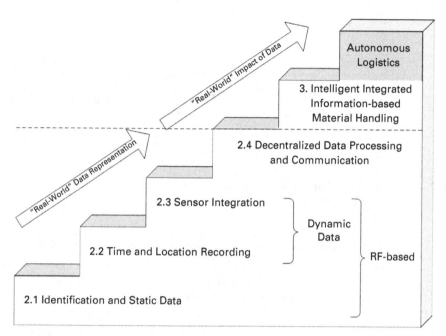

Fig. 14.1   Stepping from RFID to autonomous logistics.

intelligent are explained in Section 14.2. This approach is depicted in Fig. 14.1 by the steps of "real-world" data representation. Real-world in this context is defined as information physically bound to a logistic object. Identification and static data can be stored on RFID tags. The recording of time and location is also possible with today's RFID technology and is supported by others such as the GPS and forthcoming GALILEO systems. Sensor integration is another field of interest when using RF for logistic automation; RF-based sensors for temperature and humidity are already available. Distributed stored information, static as well as dynamic, is used within data processing and at communication level. Simple data processing may also be seen on an RF-based tag, while more sophisticated decentralized data processing is enabled with software agents.

These technologies are virtual representations of the real world, whereas in Section 14.3 we will integrate the aspect of intelligent integrated information-based material handling systems into autonomous logistics. The combination of intelligent objects and intelligent integrated information-based handling systems will allow autonomous cooperating material handling.

14.2   Radio frequency – key technology for autonomous logistics

As already mentioned, RFID is a major enabler for the emergence of autonomous cooperating logistic processes. An RFID system consists of a transponder and a

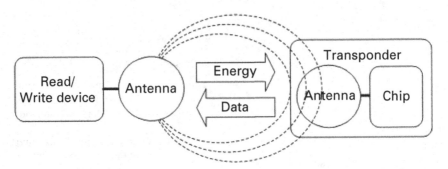

Fig. 14.2 Components of an RFID system [23].

read/write device. A transponder is a mobile data memory, which consists of a memory chip and an antenna [7]. The read/write device in a passive RFID system transfers the energy which is necessary for communication with the transponder and receives data from or writes data to the transponder (see Fig. 14.2)

Like read/write devices, transponders differ strongly from each other concerning design and abilities. RFID labels or smart labels are attached, labeled for example, to pallets, parcels, cardboard packaging or individual products. In addition, transponders can be included within the product material like a construction unit so that they merge with the article. The simplest transponders are read-only, i.e. they convey the stored information when they enter the proximity of a reader. Other transponders are re-writable, meaning that new information on the marked article can be stored and modified. Furthermore, transponders differ regarding their read range or their susceptibility to noisy influences. In general, RFID offers profound advantages over other forms of identification technologies, such as

- bulk reading,
- no line-of-sight requirement,
- robustness,
- larger memory than barcode,
- re-writable/changeable data storage, and
- ease of use.

In certain instances it might offer swifter data entry when bulk reading is possible. Even though there is an ongoing discussion on tag prices, RFID might even be cheaper than other identification solutions in certain cases.

Some of the criteria for the correct choice of RFID systems are represented in Table 14.1.

Not all of the specified criteria in Table 14.1 can be randomly combined. This is due on the one hand to technical reasons, and on the other to the fact that there are still no appropriate products available for certain combinations despite the technical feasibility.

Table 14.1. Selection criteria of RFID systems [8]

Selection Criteria for RFID Systems					
Read range	Short		Middle		Long
Transponder energy supply	Passive		Semi-active		Active
Vulnerable to water	Yes			No	
Vulnerable to metal	Yes			No	
Data memory technology	Read only		Read/write		WORM (write once/read many)
Size of data memory	≤96 bit		>96 bit; <2 kbit		≥2 kbit
Optional transponder functionality	Sensor integration (e.g. temperature, humidity)		Real-time location		Processing capability
Transponder design	Label (smart label)	Card (smart card)	Coin		Special shapes
Transponder fastening	Insertion (e.g. gusseted wallet)	Adherence/ self-adhesive	Screwing		Miscellaneous
Read/write device (handling)	Hand-held		Movable system		Stationary system
Read/write device (design)	Self-contained reader/writer	Reader/writer with data processing capability	RFID-label/card printer		Special shapes
Reading	Manual		Semi-automatic		Automatic
IT connection	Cable			Cordless	
	RS 232 C	Ethernet	Bus system	WLAN	Bluetooth	Other
Antenna	Integrated, single antenna		Separate, single antenna		Separate, combination of antennas

Identification and transmission of static data

Today, RFID technology is commonly used for identifying objects in all kinds of logistic processes. The main focus is on identifying objects at different stages throughout the supply chain. Therefore, only an ID is often stored on the transponder so that only a small memory size is required. This development or trend to only store identification data (e.g. the electronic product code, EPC) on the tag was encouraged mainly by GS1 and the Auto-ID Labs.

The ability of RFID to store data offers the possibility of using a tag as a means for offline data exchange, which makes RFID a key technology for autonomous logistics with decentralized data storage and processing. To accomplish this vision, however, transponders with a larger data memory, to store more than just the ID, are needed. The arising coalescence of material and information flow, which is a main focus of autonomous cooperating logistic processes, is advantageous where online access may be slow or difficult to accomplish.

Apart from the autonomous control approach, there seems to be a demand for tags offering more storage capability, and not only for static data. Therefore, numerous chip manufacturers are already offering corresponding EPC Generation II-compliant tags with extra user memory or have announced that such products will be available soon. While most manufacturers offer a maximum memory size of up to 1 kbit, the maximum memory available today is 64 kbytes. The additional user memory is often particular to specific branch needs. For example, the pharmaceutical industry might store lot numbers and expiration dates. Within manufacturing, extra memory may be used to establish a lot, batch or product genealogy [9]. The airline industry uses the extra memory to store lifecycle data such as equipment maintenance and inspection data. In general, there is a demand for extra user memory to

- utilize existing company- or branch-specific identification numbering schemes,
- store additional static data,
- add and change data throughout the logistic process (e.g. path and routing information),
- add and change data throughout the product lifecycle (e.g. recycling data, revision levels),
- collect sensor data (e.g. temperature, humidity), and
- store processed data (e.g. throughput data, remaining shelf-life).

The demand for tags with extra memory is not limited to UHF frequency. At 13.56 MHz, RFID tags with extra memory have been available for a long time. The Department of Defense (DoD) has expressed current requirements for active RFID transponders corresponding to the ISO 18000-7 standard, thus operating at 433 MHz [10]. The DoD expects a memory size of 128 kbytes, an unobstructed read distance of about 100 m, and a battery life of 4 years.[2] Still, due to the rapid

---

[2] "DoD Seeks New Active-Tag Suppliers" (http://www.rfidjournal.com/article/articleview/2856/1/1/).

growth of applications based on the EPC GenII standard, which has been ratified by the ISO and published as ISO/IEC 18000 Part 6C, and furthermore the availability of compliant tags with extra memory, the most vivid development will most likely be related to the UHF frequency band [11].

There are some intrinsic problems within ISO/IEC 18000 Part 6C when it comes to accessing extra memory. For example, there is no notification command to tell the reader that the tag has extra memory available and there is no defined organization of the memory space. The organization of the user memory is vendor-defined and not within the scope of the standard. The need for standardization is eminent [12].

Data transmission speed may be another problem. The maximum theoretical transmission speed between tag and reader using EPC GenII-compliant tags is limited to a maximum of 640 kbits per second. In real-world scenarios lower transmission rates will be realistic. Tests at the LogDynamics Lab[3] in Bremen with a hand-held device have shown that write times of about 3 seconds for 128 bytes can be achieved using an EPC GenII-compliant tag. On the basis of these measurements it will be necessary to have a close look at the logistic processes in order to identify the right time and place for reading and writing data to the tag. In laboratory environments today, the separation of objects and slowing down of conveyor belts are necessary in order to read and write data beyond pure identification. This will not be acceptable in most of the real-world scenarios where processing speed is crucial.

For open-loop applications it is necessary to use standards for data syntax and semantics. It seems logical to use the same standards as have been used for barcodes. The semantics for static data is defined in ISO/IEC 15418 [13]. Unfortunately, these standards are limited to static data and do not offer a solution for dynamic data such as temperature values or processed data such as, for example, remaining shelf-life. An extension of these semantic definitions for dynamic data should be considered.

Time and location recording

While identification through RFID provides visibility about the "right product" and the "right quantity," the "right time" and the "right location," two more relevant variables within supply chains, may be checked using RF technologies. When identifying a product at a fixed place such as a dock door, the time and place during the identification process may be recorded as well (e.g. goods receipt, goods issue). The state-of-the-art technology for location-based systems is now the GPS and will be GALILEO in the near future. The GPS-based location systems are limited to outdoor scenarios because a line-of-sight connection is required

---

[3] The LogDynamics Lab is located at the University of Bremen and serves as a test and demonstration environment for mobile technologies such as RFID. It is part of the Bremen Research Cluster for Dynamics in Logisitics (www.logdynamics.com).

between the tagged object and the GPS satellites. For indoor and outdoor usage in limited areas such as airports or car yards, active transponders operating in higher frequency ranges offer direct real-time locating [14]. Different methods such as

- time-of-flight ranging systems,
- amplitude triangulation,
- time difference of arrival (TDOA), and
- angle of arrival (AoA),

as well as combinations of these methods, are used to achieve an accuracy of within 3 m, as described in Ch. 6.

ISO/IEC 24730 Part 1 defines a TagBlink[4] data structure including meta-data structures for location, angle, and distance from the reader and allows two air interfaces at 433 MHz and at 2.4 GHz [15]. To provide a cost-effective means of location tracking, it is possible to combine the GPS and passive RFID in a hybrid personal data terminal as a means for time and location recording [12].

Sensor integration

To gain further dynamic data about the state of the product, sensors may be used within logistics and production scenarios to assure the "right condition" of products. Today there are sensors available to measure temperature, humidity, shock, gas concentration, acceleration, and numerous other environmental data. If wired connections to a processing unit are not applicable, RF offers a solution to communicate the sensed data to an RF-reader infrastructure.

The RF-based sensors may be categorized into passive, semi-active, and active sensors. Passive sensors will be able to record values only when activated through a reader. Semi-active sensor tags are battery-assisted and record and store values, while communication with the reader is based on passive technology. Active sensors will use their battery to record and store values as well as transmit data to the corresponding reader. An amendment to ISO/IEC 18000-6 with a corres-ponding command set for sensors is under discussion [11].

Sensing environmental influences such as humidity and temperature may be done to comply with legal requirements and is the foundation for product-specific quality models to calculate values such as remaining shelf-life. A successful inte-gration of RFID, sensor networks, and quality models into the "intelligent con-tainer" has shown the applicability of this concept to real-world scenarios [16].

Decentralized data processing and communication

There are numerous ways of processing data acquired through automated RF-based data capturing. Since there is a huge investment in existing IT

---

[4] ISO/IEC 24710-1 tag blink is defined as a "series of transmissions communicating a single asset movement or status change."

infrastructure, most approaches focus on middleware products, thus integrating new technologies and existing centralized legacy systems.

Within supply chain logistics and production scenarios, data required for efficient planning and controlling will not always be available centrally. Beside the technical burden of implementing the infrastructure to exchange information throughout the logistic process, there may be a reluctance of the business partners in supply chains to provide a complete set of business information to a centralized infrastructure. The reluctance to allocate information for open usage in supply chains is one of the reasons why the "Internet of Things" has not seen a wider adoption in open supply chains up to now [17].

Decentralized data processing may be an alternative to overcome these problems. Data processing could start on the tag itself. "Pre-processing labels" combine data storage and microprocessor functionality [18]. Owing to the limited processing capacity of these intelligent labels, data processing is bound to small calculations such as computing throughput time in manufacturing. The Collaborative Research Centre 637 document *Autonomous Cooperating Logistic Processes – A Paradigm Shift and its Limitations* takes another direction to decentralized data processing. The smallest controlling entity in this approach is a software agent, which could run on intelligent RFID readers, on personal digital assistants (PDAs) or on other hardware within a network. An agent is described as anything that is able to "perceive its environment through sensors and act upon that environment through actuators" [19]. The system architecture, a multi-agent system (MAS), is one that consists of a number of agents, which interact with one another. A single agent within the MAS is considered to be "intelligent" and "deliberative." Intelligence in this context is defined as having an internal model of the world and being able to infer on its sensor input with respect to the world model. Deliberative stands for the behavior of an agent being guided not only by stimulus–response rules but also by reasoning based on possibly conflicting higher-level goals and the world model. In real-world scenarios every packing lot (e.g. a box containing fish) would be an agent, as would the truck, the cooling unit, the warehouse, and all the other entities of the supply chain [3].

In production, logistics software agents are used to represent RFID readers, routing gates, robots, products, and individual production orders [20]. In future, the combination of RF and software agents will be a means to add intelligence to logistic objects and production facilities.

In order to realize the idea of autonomous cooperating logistics, further research concerning RF-based technologies in combination with software agents has to be conducted. Radio frequency may be used for transmitting data over short distances of up to about 10 m. Active transponders may even cover ranges of several hundred meters. However, RF today is still limited to communication between tags and special readers. For communicating data to other infrastructures beyond RF-enabled systems, other technologies such as wireless LAN, GSM/ GRPS, UMTS, and satellite communication are used. While these technologies are already widely adopted, there still remains a need for transparent switching between them. For this purpose, a communication service module (CSM) was developed at the University of Bremen within the scope of Collaborative Research Center 637. This module provides access to wireless technologies such as wireless LAN, GSM/GRPS, UMTS, and satellite communication, without the need to configure the network communication individually. Moreover, the CSM automatically chooses the right communication technology, depending on network availability and an agent-chosen profile, considering its own preferences, cost limits, quality of service, power consumption, and security [21].

In this section we have described aspects of data representation and the opportunity to combine RF-based technologies, advanced decentralized data processing, and intelligent communication infrastructure to realize autonomous logistic processes. While standards for RFID have been defined by GS1 and the ISO, there is still a need for standards when looking at memory usage and data semantics. Standardized data communication to the outside world is a prerequisite for RFID-aware automated handling systems, which will be explained in the following section.

14.3   RFID-aware automated handling systems – the differentiator between intelligent objects and autonomous logistics

To implement autonomy into logistic processes, the sophisticated intelligent logistic objects need advanced material handling systems with an ability to act accordingly. In this area there has been an evolution during the last few decades. This development of material handling systems can be categorized into six steps [22].

Step 1. Manual material handling systems. In this step material handling is conducted manually.

Step 2. Mechanical material handling systems. The material handling is supported by a forklift, conveyor belts, etc.

Step 3. Automated material handling systems. The handling is supported by mechanical systems such as automatic storage and retrieval systems (AS/RSs), automatic guided vehicles (AGVs) or robots and automated by means of electronic devices such as electronic eyes and barcode scanners.

Step 4. Integrated material handling systems. Different automatic material handling devices are coordinated and steered by a single centralized controller.

Step 5. Intelligent material handling systems. These systems are able to perform complicated tasks such as the automatic loading or unloading of trucks, steering processes in production and distribution, and planning the available material flow while using natural and artificial intelligence.

Step 6. Intelligent integrated information-based ("I^3") material handling systems. These systems based on ubiquitous and affordable information require a new model for the relationship between the different stake holders within a supply chain and flexible material handling systems to adapt to frequent changes while decreasing the throughput and response time.

Today, a mixture of these six material handling steps can be seen throughout the market. Examples of step 6 – intelligent integrated information-based handling systems – are still rare, but will be seen more frequently in the future, and will be based on RF technologies. There will be more I^3 material handling systems defined by distributed and information-connected handling systems throughout the supply chain, sharing data online (e.g. via the internet) and offline as well as performing decentralized data processing.

A logical determination for handling systems within the supply chain is to separate inbound and outbound supply chains. The inbound supply chain is affected by material, bought-in parts, and resources. Because of their different dimensions, weights, and molds on the item level, the integration of handling systems is not as common as it is within the outbound supply chain. The outbound supply chain is affected by end-user products. In comparison with the inbound supply chain, fewer different dimensions, weights, and geometries have to be handled. The application of fully autonomous handling systems in this case is more sufficient than within the inbound supply chain and allows an optimization of translation from item level up to container level. Still, for handling integration, the translation between the different application layers remains a challenge within the supply chain (Fig. 14.3).

The problem of palletizing can be solved by a standard palletizing robot with a pack algorithm. The parcel has to be allocated to a defined pallet automatically. RFID may be used to improve this process. Besides offering reliable and secure identification, RFID can provide information about the parcel's size and weight, its content, and the condition of the content as well as flexible routing information. Software agents can be used to process the collected information to act accordingly. With the integration of RFID and software agents, a fully autonomous handling of logistic objects will be possible. Autonomous handling leads to an increase in capacity and contributes to a more economical total system. The integration of RFID in material and stock flow leads to quality improvements and optimized planning of all processing steps.

There are numerous possibilities for implementing I^3 material handling systems. As an example, we focus on parcel logistics as one area of interest in this chapter. The most important environmental developments within parcel logistics are the growing standardization of shipments and technical developments. Shipments with homogeneous contents defined by a similar geometry are particularly caused by the demand of the online mail order business (B2C) as well as by the dislocation and distribution of the production facilities because of increasing globalization. The technical development plays a decisive role because the level of automation continues to rise. To meet customer demands such as reliability and speed, parcel service providers enhance their logistic centers by innovative technical solutions. Innovation is the main driving force for economic development and growth. For a logistics service provider it can be the central factor for success. If a company loses the ability to grow and shows no market improvements, it will

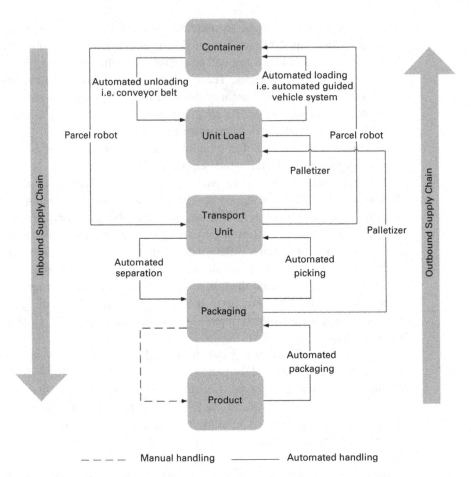

Fig. 14.3   Material handling systems within an inbound and outbound supply chain [23].

forfeit its means of existence. RFID technology in combination with robotics is one possibility allowing companies to face these demands appropriately [23]. The high quality and the speed of collecting data ensure an improved data quality for logistic planning processes. Nowadays, robots are used to palletize, staple or choose processes in logistics. At present they are hard to find in the courier express parcel (CEP) market since the unloading and loading situation is mostly characterized by chaotically stored, loosely packaged goods in swap bodies, rollaway bins, and truck-loading spaces. Up to now there has been a vacancy for a technical solution to automate this process.

An even bigger potential for unloading ceiling-high overseas containers could be verified through accomplished market studies and qualitative expert interviews with the main focus on "contract logistics." In comparison with the CEP market, the containers imported from east Asia show an even more typical dense-packing situation with homogeneously packaged goods. Next to consideration of technical demands on the functionality of the robot system, analyses of the basic conditions for automated receiving of goods have been conducted. Automatic unloading using the "parcel robot" demonstrates the first step in the process [24]. In the second downstreaming process the palletizing and picking will be an even bigger challenge to the logistic processes of the goods-receiving area.

14.4 Conclusion

Radio frequency provides the technological basis for ubiquitous identification, offline data exchange, and time, location, and environment sensing. Advanced methods of data processing such as the use of pre-processing labels and software agents as well as transparent communication hardware will help to overcome given limits of centralized data processing infrastructure. Intelligent handling systems will be able to utilize additional information stored on the RF tag to dynamically route and palletize logistic objects according to their individual requirements. Thus, autonomous cooperating logistic processes and handling will be possible. Still, some issues such as standardized data semantics and read/write cycles need to be solved in order to achieve the clock cycles required by modern logistic service providers.

14.5 References

[1] **Suh, N.**, "Complexity in Engineering," *Annals of the CIRP*, 2:581–598 (2005).

[2] **Larsen, E.**, **Morecroft, J.**, and **Thomsen, J.**, "Complex Behaviour in a Production–Distribution Model," *European Journal of Operations Research*, 1:61–74 (1999).

[3] **Chapman, S.**, *Fundamentals of Production Planning and Control* (Prentice-Hall, Englewood Cliffs, NJ, 2005).

[4] **Scholz-Reiter, B.**, **Windt, K.**, **Kolditz, J.**, **Böse, F.**, **Hildebrandt, T.**, **Philipp, T.**, and **Höhns, H.**, "New Concepts of Modelling and Evaluation of Autonomous Logistic Processes," in *Proceedings of the IFAC-MIM'04 Conference on Manufacturing, Modelling, Management and Control*, ed. G. Chryssolouris and D. Mountzis (Elsevier Science, Amsterdam, 2005) (CD-ROM).

[5] **Bemeleit, B.**, **Lorenz, M.**, **Schumacher, J.**, and **Herzog, O.**, "Risk Management for Transportation of Sensitive Goods," in *Innovations in Global Supply Chain Networks. Proceedings of the 10th International Symposium on Logistics (10th ISL)*, ed. K. Pawar, Ch. Lalwani, J. de Carvalho, and M. Moreno, pp. 492–498 (University of Nottingham, Nottingham, 2005).

[6] **Scholz-Reiter, B.**, **Windt, K.**, and **Freitag, M.**, "Autonomous Logistic Processes – New Demands and First Approaches," in *Proceedings of the 37th CIRP International Seminar on Manufacturing Systems*, ed. L. Monostori, pp. 357–362 (MTA Sztaki, Budapest, 2004).

[7] **Glover, B.**, and **Bhatt, H.**, *RFID Essentials* (O'Reilly Media, Cambridge, 2006).

[8] Scholz-Reiter, B., Mattern, F., Uckelmann, D., Hinrichs, U., and Gorldt, C., *RFID wird erwachsen – Deutschland sollte die Potenziale der elektronischen Identifikation nutzen* (Fraunhofer IRB Verlag, Stuttgart, 2006).

[9] Uckelmann, D., and Böse, F., "Von der Chargenverfolgung zur Produktverfolgung – Veränderungen in der logistischen Ruckverfolgung auf Basis innovativer Identifikationstechnologien," in *Chargenverfolgung – Möglichkeiten, Grenzen und Anwendungsgebiete*, ed. C. Engelhardt-Nowitzki and E. Lackner, pp. 133–148 (DUV-Verlag, Wiesbaden, 2006).

[10] International Organization for Standardization (ISO), *Information Technology – Radio Frequency Identification for Item Management – Part 7: Parameters for Active Air Interface Communications at 433 MHz*, ISO/IEC 18000-7:2004 (ISO, Geneva, 2005).

[11] International Organization for Standardization (ISO), *Information Technology – Radio Frequency Identification for Item Management – Part 6: Parameters for Air Interface Communications at 860 MHz to 960 MHz*, ISO/IEC 18000-6:2004 (ISO, Geneva, 2004).

[12] Harmon, C., "The Necessity for a Uniform Organisation of User Memory in RFID," *International Journal of Radio Frequency Identification Technology and Applications*, 1:41–51 (2006).

[13] International Organization for Standardization (ISO), *Information Technology – EAN/ UCC Application Identifiers and Fact Data Identifiers and Maintenance*, ISO/IEC 15418:1999 (ISO, Geneva, 2005).

[14] Scholz-Reiter, B., and Uckelmann, D., "Tracking and Tracing of Returnable Items and Pre-Finished Goods in the Automotive Supply Chain," in *Proceedings of the 1st RFID Academic Convocation*, ed. S. Miles, B. Hardgrave, and J. Williams (Boston, 2006).

[15] International Organization for Standardization (ISO), *Information Technology – Real-time Locating Systems (RTLS) – Part 1: Application Program Interface (API)*, ISO/ IEC 24730 (ISO, Geneva, 2006).

[16] Jedermann, R., Uckelmann, D., Sklorz, A., and Lang, W., "The Intelligent Container: Combining RFID with Sensor Networks, Dynamic Quality Models and Software Agents," in *Proceedings of the Second RFID Academic Convocation*, ed. S. Miles, B. Hardgrave, and J. Williams (Las Vegas, 2006).

[17] Schuster, E., Allen, S., and Brock, D., *Global RFID. The Value of the EPCglobal Network for Supply Chain Management* (Springer-Verlag, Berlin, 2007).

[18] Overmeyer, L., Nyhuis, P., Höhn, R., and Fischer, A., "Controlling in der Intralogistik mit Hilfe von Pre Processing Labels," in *Wissenschaft und Praxis im Dialog: Steuerung von Logistiksystemen – auf dem Weg zur Selbststeuerung – 3. Wissenschaftssymposium Logistik*, ed. H. Pfohl, pp. 205–215 (Deutscher Verkehrs-Verlag, Hamburg, 2006).

[19] Russell, S., and Norvig, P., *Artificial Intelligence: A Modern Approach*, 2nd edn. (Prentice-Hall, Englewood Cliffs, NJ, 2003).

[20] Mařík, V., and Vrba, P., "Simulation in Agent-Based Control Systems: MAST Case Study," in *Preprints of the 16th World Congress of the International Federation of Automatic Control*, ed. M.Šebek (IFAC, Prague, 2005) (CD-ROM).

[21] Timm-Giel, A., and Görg, C., "Self-Configuring Communication Service Module Supporting Autonomous Control of Logistic Goods," in *Proceedings of the 3rd International Workshop on Managing Ubiquitous Communications and Services (MUCS 2006)*, ed. D. Pesch, pp. 2–9 (Cork, 2006).

[22] **Pence, I.**, "A Perspective on Material Handling Engineering: History and New Challenges," *Journal of Manufacturing Science and Engineering*, 4:835–840 (1997).

[23] **Scholz-Reiter, B.**, **Uckelmann, D.**, **Gorldt, C.**, and **Hinrichs, U.**, "Einsatz von Auto-ID-Technologien in der Handhabungstechnik – Neue Entwicklungen in der Inbound/ Outbound Supply Chain," *ZWF – Zeitschrift für wirtschaftlichen Fabrikbetrieb*, 3:97–101 (2006).

[24] **Echelmeyer, W.**, **Tank, S.**, and **Wellbrock, E.**, "Einsatzmöglichkeiten von Industrierobotern in Paketverteilzentren," in *Roboter in der Intralogistik*, ed. R. Schraft and E. Westkömper, p. 123 (Verein zur Förderung produktionstechnischer Forschung, Stuttgart, 2005).

15   Conclusions

Stephen Miles, Sanjay Sarma, and John Williams

The principal investigators represented in *RFID Technology and Applications* have presented a range of practical approaches and models to consider in preparing an RFID implementation strategy as well as for planning future research areas for use of this fast-growing technology. As Sanjay Sarma states in presenting applications for RFID (Ch. 2), passive RFID technology is still in its infancy. We have identified challenging aspects of linking autonomous agents and intelligent handling systems (Ch. 14) with large volumes of distributed data sources such as RFID (Ch. 7). Using a consistent experimental approach based on control theory, test and simulation frameworks have been proposed to help evaluate RFID performance, for applications that start with uniquely identifying products and can include exchanging this information, together with sensor and real-time location data, across enterprises.

By way of summary, three areas emerge from these research initiatives that warrant careful planning, starting with the challenges of using low-power wireless data acquisition technology, with respect both to electromagnetic performance of tags in relationship to specific products and packaging, and to downstream RF environments in which the tags are to be read. These issues have been explored in the technology section (Chs. 2–8).

A second theme is the requirement to extend visibility over entire product lifecycles, an increasingly recurrent topic as governments seek new ways to gain visibility on globalizing supply chains. In this context RFID is being evaluated not just for tracking outbreaks of contaminated products but also for holding suppliers responsible for the environmental, safety, and economic (tax) impact of their product as seen in retail, aerospace, cold chains, and automotive end-of-life vehicles (ELVs) (Chs. 9–14).

This leads us to a third common theme, which is the promise of RFID and the EPC to provide a new level of visibility and granularity of information from the physical world. Because this data is not hardwired into legacy proprietary data models, auto-ID data holds the possibility of communicating about uniquely identified objects in new ways and providing actionable information based on physical events as exemplified in the autonomic logistics processes developed by BIBA Labs at the University of Bremen (Ch. 14) in their work with sponsors such as DHL.

RFID Technology and Applications, eds. Stephen B. Miles, Sanjay E. Sharma, and John R. Williams. Published by Cambridge University Press. © Cambridge University Press 2008.

Throughout the chapters of this book, RFID interface specifications, test capabilities, and simulation environments have been presented by researchers in order to facilitate planning and evaluation of RFID technology. A list of the research test and simulation environments identified, and links to associated co-authors' research initiatives, are appended at the end of this chapter. We review these capabilities with respect to how they relate to addressing key elements of an RFID strategy.

15.1 Radio frequency gap analyses; Georgia Tech LANDmark Medical Device Test Center

The selection of the frequency (or frequencies) at which RFID tags are to be read is the first engineering choice that users are faced with, whether selecting low-cost passive tags or more-expensive active transponder devices. To illustrate the extreme differences in wireless environments Marlin Mickle (Ch. 4) uses the analogies of an "RF prison cell" such as a shielded anechoic chamber versus the "bird in the sky" warehouse dock or checkout counter with very different RF noise to see through. One of the challenges of "open-loop" RFID applications is in planning for the range of conditions under which RFID tags may be read "beyond the four walls." New generations of multi-frequency readers and/or tags promise to extend the range of conditions under which tagged items can be "seen."

The case of how to test facilities with unique RF operating conditions such as hospital and military environments is introduced by Gisele Bennett and Ralph Herkert (Ch. 8). As early as during the first RFID Academic Convocation, the EPCglobal Healthcare Life Sciences (HLS) Business Action Group Co-Chairs presented the need for a "gap analysis" of RF applications specific to medical facilities. At the subsequent fifth RFID Academic Convocation held in conjunction with RFIDLive!2007 in Orlando, Ralph Herkert of the Georgia Tech Medical Device Test Center presented how their laboratory for testing interference of cell phone and EAS systems with medical devices could be used by RFID equipment manufacturers. In their chapter Gisele Bennett relates her experience working in industry-specific RF environments such as military installations, where 433-MHz RTLS tracking systems may interfere with radar systems operating within the 420–450-MHz spectrum. Gisele Bennett notes common WiFi infrastructure interference issues such as those involving legacy 900-MHz WLAN systems that have not yet been migrated to 2.45GHz and that may cause interference in the UHF 868–928-MHz bands used by EPC GenII/ISO 18006c RFID systems.

15.2 The RFID Technology Selector Tool; Auto-ID Labs at Cambridge University

In considering whether to deploy active and/or passive RFID systems at HF or UHF frequencies, using near-field or far-field RF radiation, a series of variables must be weighed against each other to assess the best technology fit. One planning tool that is being developed was presented in the chapter by Duncan McFarlane and colleagues on the Aero ID Initiative (Ch. 10). As a part of this initiative, the Auto-ID Labs at Cambridge is developing a tool to assist with RFID tag technology selection based on business and operating environment requirements, seeking to balance timeliness or read rates against accuracy and completeness. The example of active tag requirements specific to the aerospace industry as specified in the ATA2000 aircraft industry specifications includes requirements to read tags on metals over long durations and to store maintenance information on tags. A different example of active RFID sensor applications from the "cold chain" industry is provided by J. P. Emond (Ch. 11), from the Center for Food Distribution and Retailing at the University of Florida in Gainesville, where testing of transponders with temperature sensors embedded in the packaging is under way. The physics of achieving high read rates in densely packed metallic containers containing liquids is one of the challenges that OEMs are hoping to address with new near-field electromagnetics as described by Sanjay Sarma (Ch. 2) and by Marlin Mickle and colleagues (Ch. 4). In Hao Min's exposition of a low-power RFID tag design we saw key characteristics for optimizing chip performance. Hao Min's overview of a verification and test methodology is a good illustration of how control theory can be applied to optimizing tags for specific products and/or operating environments (Ch. 3).

15.3   An EPC GenII-certified test laboratory; the RFID Research Center, University of Arkansas

Manufacturers who wish to test the performance of tags on their products downstream in the supply chain have the option of working with EPCglobal-certified test laboratories, one of the first of which was established by Bill Hardgrave, Executive Director of the Information Technology Research Institute at the University of Arkansas and Co-chair of the RFID Academic Convocation. The geographic proximity and close working relationship that Bill Hardgrave has established with the largest retailer in the world make this laboratory an excellent choice for RF testing scenarios that involve replicating Wal-Mart's distribution center RF environments, examples of data transmissions from which he analyses with Robert Miller, Ph.D., from Ashland University in Ohio in their chapter (Ch. 9)

15.4   ISO 18000-7 and 6c (HF and UHF) RFID and EPC network simulation

In pursuing a methodology for RFID systems testing, specific issues have been identified when testing RFID systems performance where vendor implementations differ. Simulations can help address questions about how frequency choices will impact modulation and transmission schemes as well as coding trade-offs at the MAC layer of the RF communications medium. A layered approach to RFID reader protocol testing (following the ISO protocol stack) is proposed by Marlin Mickle, who calls for revising RFID protocol specifications accordingly (Ch. 4).

The HF/UHF and near-field/far-field simulation environment; RFID Center of Excellence, University of Pittsburgh

In the absence of a layered RF protocol specification that monitors connectivity at the PHY layer prior to engaging MAC-layer communications, Marlin Mickle cautions, we cannot accurately describe "RFID read rate" and "write time" performance. As a result project managers must identify ways to consistently define reader performance according to specific operating characteristics in order to successfully measure outcomes. For example, it has been shown that generic UHF readers that have been modified to support HF frequencies using a UHF protocol do not generally perform as well as native HF readers. In describing the evaluation process for what frequencies to use for case- and/or item-level tagging, Bill Hardgrave and Robert Miller cite concerns at large RFID installations where HF read rates can be up to five times slower than UHF (Ch. 9). Similar concerns are expressed in the evaluation by Duncan McFarlane and co-authors in the Aero ID Initiative at Cambridge University (Ch. 10). Bernd Scholz-Reiter, Dieter Ucklemann, and co-authors (Ch. 14) express the same concerns as per tests at the LogDynamics Lab in Bremen, Germany, in which a hand-held UHF device exhibited write times of about 3 seconds for 128 bytes using an EPC Gen II-compliant tag. Strategies for using a single transponder, one reader with multiple antennas, or perhaps a cluster of readers or even a sequence of readers installed throughout a facility, as described in Sanjay Sarma's introduction to RFID applications (Ch. 2), constitute a further area requiring analysis.

To isolate each layer of RFID read/write performance the University of Pittsburgh has produced a development suite that enables rapid design and implementation of readers and tags compliant with ISO 18000 Part 7 and is close to completion with a similar system that would do the same for readers and tags compliant with ISO 18000 Part 6C. Marlin Mickle asserts the value of this analytical approach in his concluding remarks for Ch. 4: "Rather than using an RFID reader as the basis for drawing conclusions as to systems performance, you really need to analyze the RF exchanges to see what is actually happening. Transmitter power, receiver sensitivity, and noise are the primary variables to measure, along with processor and memory size on the tags." Furthermore, Marlin Mickle proposes a layered approach to RFID protocols that could provide the basis for frequency-independent protocol testers that would analyze RFID performance in specific environments, beginning with whether or not interrogators have established physical-layer wireless connectivity. Such a project will likely require sponsored research collaboration with multiple RFID suppliers as well as academic partners.

The EPCIS Accada Java open-source initiative and RFID reader simulator

Christian Floerkemeier of the Auto-ID Labs at MIT and the ETH Zürich has developed, with project coordinators Christof Roduner and Matthias Lampe of the Auto-ID Labs at St. Gallen and the ETH Zürich, an open-source Java version of EPCIS [1] that includes a reader-simulation capability to emulate the performance of large numbers of tags and readers as displayed in Fig. 15.1.

To facilitate testing of RFID applications, the built-in simulation framework includes a graphical user interface that allows the developer to drag and drop tags over different reader antennas. The simulator also provides a mechanism to schedule the detection of a tag at different times on different reader instances. This allows a developer to simulate the movement of a tag population through the supply chain.

An EPC network simulator and EPCIS Reference Code; MIT Auto-ID Labs

One approach to ensuring that the various EPC network systems being proposed in the standards arenas can work together is to build models to simulate actual performance, as described by John Williams, Abel Sanchez, Paul Hoffman and colleagues (Ch. 7). Security and privacy go hand in hand, especially in the case of some products, such as pharmaceuticals, regarding which abiding by legal requirements for anonymity is a prerequisite to information sharing. Companies, like individuals, do not want to give up the ownership of their data without agreements in place to limit access and to ensure that they will share the benefits of such an exchange. In investigating how to share data securely, John Williams discusses models for registries and security credentials to complement EPCglobal specifications for the use of X.509 certificates that would authenticate third-party inquiries in a multi-tier

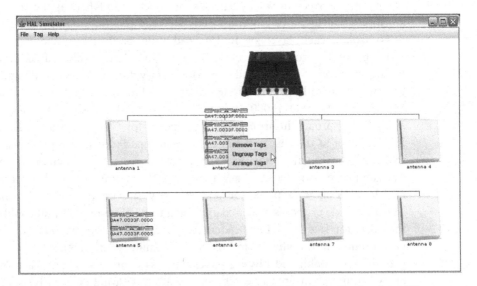

Fig. 15.1   A screenshot of the Accada open-source Java EPCIS-based reader simulator [1].

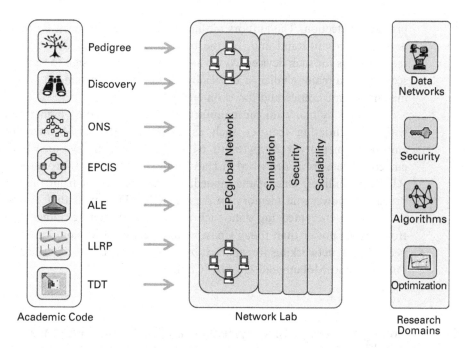

Fig. 15.2 The MIT Auto-ID Labs open-source .NET EPCIS implementation (MENTOR).

supply chain such as in healthcare life sciences, as is being attempted in the EPC network simulator project with SAP Research.

An integral part of the EPC network simulator has been the development of a complete MIT Auto-ID Labs open-source EPCIS reference code base in .NET, as illustrated in Fig. 15.2. The MIT EPC net (MENTOR) has been developed under the leadership of John Williams, Principal Investigator, and Abel Sanchez, Research Scientist, to provide demonstrators, documentation, and an EPCIS interoperability reference code base for the research community [2].

In MENTOR, the Auto-ID Lab at MIT is introducing a software platform for RFID and sensor networks. According to Abel Sanchez, "the open-source project is based on an 'architecture of participation.' From the outset, our goal has been an architecture in which users pursuing their own self-interests will build collective value as an automatic byproduct. In that spirit, we invite the research community and industry to participate in the MENTOR platform."

15.5 RFID anti-counterfeiting attack models; Auto-ID Labs at St. Gallen and the ETH Zürich

Thorsten Staake, Florian Michahelles, and Elgar Fleisch discuss in their work on anti-counterfeiting strategies (Ch. 12) the application of RFID for brand and product protection. The authors outline attack scenarios which RFID transponders have to withstand and, on the basis of this analysis, derive the technical requirements for such devices. As we have seen, systems design to resist RF attack scenarios is a major challenge, especially for low-cost transponders with restricted maximum gate counts and limits on energy consumption.

The trend by many silicon manufacturers toward creating higher-margin tags with differentiating features such as more powerful FSM microcontrollers, embedded memory (128 bytes to 64 kbytes of EEPROM or FLASH), and embedded sensors as described in Hao Min's chapter (Ch. 3) promises to expand RFID capabilities. One of the immediate areas to benefit from the addition of processing power and memory on these new RFID chips is that of applications that require encrypted tag data such as anti-counterfeiting. Other beneficiaries from higher power than today's passive backscatter technology provides include peer-to-peer networking for applications such as RTLS as described by Kaveh Pahlavan and Mohammad Heidari (Ch. 6).

15.6   Adding sensors to RFID Systems – IEEE 1451/NIST interface specifications

The IEEE 1451 specifications introduced by Kang Lee and Tom Cain (Ch. 5) propose a basic set of transport-independent sensor messages and support widely used common wired and wireless interfaces for sensors-to-network node communications – including Enhanced SPI, UART, Ethernet (IEEE 802.3), WiFi (IEEE 802.11 a,b,g,n WLAN), Bluetooth, CANBus, and ZigBee RFID (IEEE 802.15.4 WPAN). One area for future research would be on EPC GenII specifications to support for IEEE 1451 – giving sensors the possibility of communicating without on-chip power. For active RFID sensors, another goal of standardizing the physical and logical interface would be to support a common user memory configuration and to enable web-based monitoring.

Since the topic of sensor interface specifications was introduced by Kang Lee of the National Institute of Standards and Technology (NIST) at the first RFID Academic Convocation hosted by MIT in January of 2006, standards bodies including the ISO and industry groups including the DoD, MIMOSA,[1] and OA&M, as well as leading manufacturers, have introduced specifications and products that support the self-identification of sensors on the basis of

[1] MIMOSA – the Manufacturing Interoperability Guideline Working Group – was formed in March of 2006 by a collaborative venture consisting of the ISA, the Open Applications Group (OAGi), the OPC Foundation, and the World Batch Forum (WBF). Subsequently the US Department of Defense mandated the use of these standards. MIMOSA includes support for CMMS (computerized maintenance management system); EAM (enterprise asset management); STEP (standard for exchange of product model data); OPC (OLE for process control); and OAG (open applications group). It has adopted the IEEE 1451 specifications for incorporating dynamic sensor data input into condition-based monitoring applications.

IEEE 1451.4.[2] Most recently IEEE 1451 specifications have been included as an amendment to ISO SC31 24753 specifications "Radio Frequency Identification (RFID) for item management – Air Interface Commands for Battery Assist and Sensor Functionality." In addition, ISO 18000-6(E) Rev1 integrates IEEE 1451 into the ISO18000-6 framework as a transport-independent set of common sensor commands based on IEEE 1451.0.

These technology standards are designed for applications such as condition-based monitoring for common machine tools, including mechanical equipment such as spindle motors, axis drive motors, and bearings in applications from aerospace to caterpillar tractors, as illustrated in the chapters covering European Commission Bridge projects (Chs. 10, 12, and 13). Kang Lee, in his concluding remarks, says that "currently I am working with the DHS standards office to see how IEEE 1451 smart sensor interfaces can be used for chemical, biological, radiological, and nuclear (CBRN) sensor and detector connection to systems and networks that can be deployed nationwide through the state and local governments. I am doing this through the Sensor Standards Harmonization Working Group that I chair at the NIST for the DHS. The Open Geospatial Consortium has adopted 1451 as its infrastructure base for getting sensor inputs." It is worth noting that the OGC initiative may address the requirement identified by Bernd Scholz-Reiter, Dieter Uckelmann, and colleagues from the University of Bremen (Ch. 14) for a standard specification for interfaces to dynamic sensor data for that community of users.

15.7   Adding location interfaces

While most organizations will not wish to exchange data with specific geo-positional content freely, the concept of tracking items in space is relevant to applications in transportation logistics, warehouse automation, and manufacturing process control. Adding positioning technology best practices to the portfolio of RFID-related communications is an area where the authors in this book break new ground, as proposed by Kaveh Pahlavan and Mohammad Heidari (Ch. 6). Their insights are validated in chapters describing applications for real time location systems (RTLSs) in smart containers, cold chain management, aerospace, and product lifecycle management (PLM) (Chs. 10, 12, and 13). The test methodology outlined in Ch. 5 to

---

[2] IEEE 1451.4 directs its attention to the TEDS part of the sensor and signal conditioning system alone. It adopts a valuable approach by taking a much simpler approach to other smart sensor concepts by simply focusing on the self-identification aspects of a sensor. It does this by specifying a table of self-identifying parameters that are stored in the sensor in the form of a transducer electronic data sheet (TEDS). National Instruments and its sensor partners are suggesting that sensor vendors start producing sensors based on the draft specification. This includes companies such as Honeywell Sensing and Control/Sensotech, one of the world's largest single-source suppliers of sensors and switches (http://www.sensotec.com/pnpterms.shtml#a9), and Maxim's 1-Wire serial interface sensors (http://www.maxim-ic.com/appnotes.cfm/appnote_number/2965).

evaluate RTLS performance provides a framework for project managers to evaluate RTLS technologies in conjunction with RFID system and frequency choices for the respective infrastructures.

Recent developments in the availability of RAND IPR licensing terms for 2.4-GHz RTLS developed by Wherenet as standardized in ISO 24730 and in 433-MHz e-seal container-tracking technology from SAVI have been standardized both in ISO 18185 and in the 18000-7 Air Protocol. Bernd Scholz-Reiter, Dieter Uckelmann, and colleagues from the LogDynamics Lab at the University of Bremen (Ch. 14) describe ISO 24730 Part 1 as a TagBlink data structure including meta-data structures for location, angle, and distance from the reader that allows air interfaces both at 433 MHz and at 2.4 GHz. The draft of ISO 24730 defines message formats between an application server and a localization server for describing location data using a SOAP interface.[3] Nonetheless, as Kaveh Pahlavan describes (Ch. 6), in the case of proprietary RTLS protocols such as are used by Ekahau, even if the transport protocol selected is an accepted standard such as 802.11g, use of a proprietary UDP/IP protocol to communicate between the active tags and APs would preempt the use of these ISO protocols for RTLS messaging with their system.

The ability to associate things that are being tracked with location reference information, whether via WiFi in-building positioning active RFID tags or via GPS receivers that may be mounted on forklifts in the yard, for example, or via wireless telephony mobile positioning systems, is one area with short-term payback through better asset management. Unlike early predictions whereby every warehouse shelf would require dedicated RFID infrastructure, RFID readers in many instances are being deployed today in clusters for high-speed manufacturing, picking and packing, shipping and receiving, and/or as mobile devices on lift trucks or as hand-helds. Where in specific applications smart shelves have value, especially for high-value items such as for tracking defibrillators or stents in hospitals, initial RFID implementations often start with the purchase of mobile RFID readers.

One research area for merging RTLS data with product information in retail supply chains concerns marrying the location of mobile readers to the geo-spatial context of new warehouse automation/robotics systems. High-speed automated picking and packaging systems are used in pharmaceutical manufacturing and distribution facilities that increasingly rely on robots to automate processing. Hutchinson ports, in their joint venture with SAVI Networks, are in the midst of upgrading port automation infrastructure around the globe to track shipping container locations within the shipping yard. Future applications that further reduce the number of readers required to cover large surface areas may evolve

---

[3] According to Tim Harrington, Vice President of Product Strategy at Wherenet, the 433 MHz vendors specified an RTLS API for ISO 24730-1 with a SOAP interface. It was designed during the INCITS ANSI process, and included inputs from several competitors, including RF-Code, RF Technologies, Bluesoft (now AeroScout), I-Ray, and SAVI. It includes ways to describe location in terms of X–Y–Z, distance plus angle, etc.

from the flying RFID robot research project at MIT Auto-ID Labs, where JinHock Ong, EECS Master of Engineering, is integrating an RFID reader to mount with an unmanned aerial vehicle (UAV) in cooperation with the Aerospace Controls Laboratory at MIT.[4]

15.8 Convergence of RFID infrastructure: multi-frequency and multi-protocol

By the spring of 2007, at events including the EU RFID Forum/RFID Academic Convocation in Brussels it had become apparent that distinctions between RFID systems operating at different frequencies and between active and passive tags are blurring as vendors design systems that support multi-frequency and multi-protocol devices. These include multi-frequency HF/UHF RFID readers from such companies as AWID, ThingMagic, SAMSys, and Tagsys as well as the Hitachi EPC GenII "Lite" UHF/2.4-GHz reader for higher frequencies, and innovations at low frequencies such as the IEEE P1902.1 "Rubee" initiative proposed by John Stevens of Visible Assets and licensed by Epson [3]. Within a given frequency, near-field capabilities are being included with far field, as exhibited by Impinj. Simultaneously, as has been discussed, RFID tags are being equipped with varying capabilities, including extended memory for encryption and/or industry-specific requirements such as with the Intelleflex product designed to meet ASA2000 specifications for recording aircraft maintenance and repair data as introduced in the Aero ID Initiative (Ch. 10). Toppan Forms "crystal on a chip" design will furthermore support multiple RFID frequencies on the RFID tags themselves [4].

At the event in Brussels, Ryo Imura, General Manager of Hitachi, Ltd., and Professor, University of Tokyo, presented Hitachi's tiny (154 mm × 13 mm) version of an ISO 18000-6c "lite" compatible inlay (chip+) with a target price in volume of 5¢, together with a dual-mode reader for both 865–7-MHz UHF and 2.4-GHz frequencies. The Hitachi μ-Chip was initially targeted at being embedded in paper currency as an anti-counterfeiting measure. The admission ticket system for the 2005 World Exposition, which had approximately 22,050,000 visitors, employed the μ-Chip, had a performance record of no incidence of confirmed forgery and 0.001% incidence of ticket-recognition error. Another UHF protocol targeting this market is NFC, according to Edward Gonsalves of NXP North America, who gave a presentation at the MIT Enterprise Forum RFID SIG on "RFID in Sports"[5] describing the successful processing of 3.2 million tickets

---

[4] See the UAV SWARM Health Management Project website, of the Aerospace Controls Laboratory at MIT (http://vertol.mit.edu/) and that of the robot team at MIT Auto-ID Labs (http://autoid.mit.edu/CS/photos/robots/picture27779.aspx).

[5] See http://www.mitforumcambridge.org/rfid/events/2007-06/.

across multiple stadiums at the FIFA World Cup for Soccer and included photos of his home country's soccer team from Portugal to boot.

In the converging RFID platforms we envisage, in addition to multi-frequency and peer-to-peer networking applications for active RFID devices, that personal communicators such as the phone and the PDA will become the readers/interrogators of choice. As Serge Ferre, Vice President of Nokia (BE), stated at the Brussels RFID Academic Convocation, there are already 14 antennas built into the typical cell phones that have millions of users and operate both at HF and at UHF frequencies for a variety of applications, including most recently the addition of NFC near-field capabilities such as in the FIFA ticketing application described above.

15.9   New business processes: from e-Pedigree to VAT tax compliance

In response to globalizing trade, government agencies in many countries are increasingly counting on improvements in AIDC/auto-ID technology to hold manufacturers accountable, both in terms of product safety and from a tax liability standpoint. One precedent that manufacturers may want to consider is the cooperation agreement between the European Commission and Philip Morris International, Inc. (a subsidiary of the Altria Group) that was signed in July of 2004 [5]. In this agreement PMI agreed to make a $125 billion settlement against VAT taxes from "gray market" shipments in exchange for a broad cooperation with European law enforcement agencies on anti-contraband and anti-counterfeit efforts. Jim Nobel, CIO of the Altria Group and sponsor of MIT Auto-ID Network Research, in his speech at RFID Live! in March of 2006 noted that item-level tagging for anti-counterfeiting across Phillip Morris and Kraft Foods product lines requires on the order of 8 billion tags per annum. Jim Nobel emphasized the importance of harmonizing compliance and anti-counterfeiting reporting standards across international supply chain stakeholders and health and safety regulatory agencies around the world.[6]

The promise of RFID applications described in this book represents game-changing benefits when coupled with new business processes that help users make sense of the data and to take action, as, for example, in the PMI case, when a customs agent interdicts counterfeit merchandise at the border by using PMI code verification system (CVS) product authentication over the network. Dimitris Kiritsis and colleagues develop a networked system for tracking ELVs using

---

[6] "In July 2004, PMI entered into an agreement with the European Commission (acting on behalf of the European Community) [...] that provides for broad cooperation with European law enforcement agencies on anti-contraband and anti-counterfeit efforts [...] In the second quarter of 2004, PMI recorded a pre-tax charge of $250 million for the initial payment. The agreement calls for payments of approximately $150 million on the first anniversary [...], $100 million on the second anniversary, and $75 million each year thereafter for 10 years."

product lifecycle management (PLM) (Ch. 13) to assist in recycling the source of 10% of hazardous waste in landfills across Europe.

Returning to RFID's roots in automatic identification technology (AIT), beginning with barcodes, identification standards have enabled tremendous growth in global commerce over the last decades. As Bernd Scholz-Reiter, Dieter Uckelmann, Christian Gorldt and colleagues conclude (Ch. 14), new decentralized and autonomous controlled processes will be needed to maximize the usefulness of this data, with business benefits resulting from cross-organizational flexibility, adaptability, and reactivity. We have seen a vision of closer collaboration through autonomic logistics (Ch. 14). Bernd Scholz-Reiter, Dieter Uckelmann, Christian Gorldt *et al.* propose a distributed system of intelligent objects on the model of a "heterarchy" wherein objects behave like neurons in responding dynamically to changing circumstances in the supply chain, and whereby intelligent integrated information-based material handling systems create autonomous logistics functions. In the Economist Intelligence Unit article "RFID Comes of Age," David Jacoby states that "RFID, and the vast volumes of data that it generates, needs to be integrated into operational management tools such as ERP (enterprise resource planning) software. Used in this way, RFID could become a catalyst for much deeper collaboration between companies, and lead to the formation of 'supplier networks' that will replace today's linear supply chains." [7]. Furthermore, the technology of RFID sensor networks, together with software agents and wireless communication, promises to extend the internet in new ways.

At the July 2006 meeting of the EPCglobal Board of Governors hosted by MIT, Dan Roos introduced the MIT lean manufacturing principles that broke with brittle centralized manufacturing models [8] as have been echoed in the Aero ID Programme presented herein (Ch. 10). For RFID automated data capture to work between aircraft subcontractors on multiple continents, specifications are essential at multiple levels, between the transponder and the interrogator, in dense interrogator environments, and in inter-networking collaboration for shared business processes. The Auto-ID Labs at MIT have documented learnings from EPCglobal data exchange work group pilots whose findings are available to EPCglobal members in the *EPCglobal RFID Cookbook* [9].

As we have seen, the challenges of processing large volumes of RFID and telemetry data are significant, as in the case of distinguishing between radar and random data described in the historical introduction to this technology (Ch. 1). A recent description of such a dilemma was given by fellow panelist, Gary Gilbert, Director of Security for Hutchinson Wampoa Ports, who asked, "What actions would you take, if you were the mayor of the port of Tacoma, in Washington state, and a container blew up outside of Long Beach harbor in California?"[7] What kind

---

[7] Gary Gilbert, CSO of Hutchison Port Holdings, Lead Panelist; Stephen Miles, Auto-ID Labs, Moderator; "Smart and Secure Trade Lanes: Demonstrating How AIDC Technologies Can Improve Homeland Security and Supply Chain Efficiency;" Frontline Conference and Exposition, Chicago (September 28, 2004) (http://www.frontlineexpo.com/frontlineexpo2004/V41/index.cvn?id=10065&p_navID=).

of processes would you design that would help your organization sift through the data from transportation and logistics systems to gain visibility on the situation?

Formulating standards for data exchange and defining shared common business processes is a labor-intensive effort that is occurring in industry associations and global standards organizations around the world. In the EPCglobal requirements definition process that the Auto-ID Labs support, business action groups have been formed for fast-moving consumer goods, healthcare life sciences, transportation logistics services, and aero defense. As was the case for the adoption of barcode standards, integration with legacy coding systems from different industries was an integral part of the process, as described by Alan Haberman in the preface to this work. Deciding how to put together data from the various unique identification systems employed by governments such as that of the USA, where asset management systems track iUIDs for material valued at $5,000 and above, while logistics information for the same material uses the EPC unique identifiers, or for pharmaceutical products, for which EPCglobal and GS1 coding schemes must now be designed to incorporate the National Association of Pharmacy Distributors (NAPD) numbering schemes are examples of the challenges faced in cross-industry collaboration. Data integration, as described by the authors from the University of Bremen (Ch. 14), and the translation between different industry identifiers remains a challenge for supply chain data exchange.

In his chapter titled "Business Navigation and Real-World Awareness," Claus E. Heinrich, member of the Executive Board of SAP AG, describes how, as the plane takes off, the airplane feeds the pilot selective information from automated sensory systems for engine speed, altitude, radar, GPS, and fuel levels that helps the pilot navigate [10]. The addition of telemetry and location information to RFID data reflects seminal changes under way in edge devices, information networks, and how the web can help us relate to the physical world. We do not have to look any further than advances in computer gaming to see how 3D visualization such as in *Second Life* and the haptic hand-held controls of Nintendo might help us to communicate with robotic picking systems or remote supply chain processes in the future. Whether a platform such as the EPC GenII readers plus sensors plus RTLS will be successful in supply chain applications has still to be determined.

New platforms engender new business models and services, which are likely to be the differentiators that create the next generation of profitable products on models similar to the Rolls Royce jet engine "power by the hour." One example of how auto-ID technologies might enable a change in business models would be to enable governments to procure and manage Medicare reimbursements as evidence-based care and pay for performance.

As the authors of this book have demonstrated, RFID technology performance and RFID applications in supply chains are non-linear problems as we move upstream into materials management and downstream to the SKUs that pass the point-of-sale terminals at retail. This book investigates how RFID item-level visibility can give visibility on supply chains that are increasingly stochastic and dynamic. Bill Hardgrave notes the importance in "open-loop" supply chains that extend beyond the supplier across DCs to the retail store (Ch. 9). In a single industry such as food distribution, Jean Pierre Emond notes that half of the $950 billion US retail food business deals with perishables (Ch. 11). One result of globalizing supply chains highlighted by Thorsten Staake, Florian Michahelles, and Elgar Fleisch (Ch. 12) is increased risks of counterfeiting and logistical vulnerabilities. A total systems view, as proposed in the Cambridge University Aero ID initiative (Ch. 10), whereby the total cost of airport delays across all carriers is the starting point for a systematic analysis, is where the highest returns on sharing RFID information are likely to be found. Their example describes how, to supplement established communication protocols for pilot-to-air traffic control point-to-point communications, a distributed system such as RFID could provide automated updates on the status of food delivery and cleaning services that then could be factored into take-off scheduling.

As applications for wireless identification of products (the electronic product code for example), of people (the "Real ID" Act, ePassports...), and of places (GLIN, DUNS, Long/Lat...) proliferate, along with expected increases in telemetry power on chips, a richer body of information than ever can be connected with things. One new way to relate disparate data sources is considered by Dimitris Kiritsis (Ch. 13) as he describes the use of RDF for linking data to the ontological reference source of origin. Building systems and services that can bring data together from these disparate sources and industries in the appropriate business process context remains a research and engineering challenge for which we welcome specific use cases and requirements input. Your participation in the ongoing RFID Academic Convocations is invited: http://autoid.mit.edu.

15.10   References

[1] **Floerkemeier, C.**, **Lampe, M.**, and **Roduner, C.**, "Facilitating RFID Development with the Accada Prototyping Platform," *4th RFID Academic Convocation/EU RFID Forum*, Brussels 2007, ed. H. Barthel, D. McFarlane and S. Miles (European Commission Directorate General for Information, Society and Media, Brussels, 2007) (http://www.rfidconvocation.eu/Papers%20presented/Technical/Facilitating%20RFID%20Development%20with%20the%20Accada%20 Prototyping%20Platform.pdf0).

[2] **Williams, J. R.**, and **Sanchez, A.**, "Supply Chain Realms with Data Streams and Location Services," *EU RFID 2007 Academic Convocation*, Brussels, ed. H. Barthel, D. McFarlane, and S. Miles (European Commission Directorate General for Information, Society and Media, Brussels, 2007) (http://www.rfidconvocation.eu/Papers%20presented/Technical/Supply%20Chain%20Realms%20with%20Data%20Streams%20and%20Location%20Services.pdf).

[3] **IEEE**, "IEEE Begins Wireless, Long-wavelength Standard for Healthcare, Retail and Livestock Visibility Networks" (2007) (http://standards.ieee.org/announcements/pr_p19021Rubee.html).

[4] "First Multifrequency Chip Unveiled; Toppan Forms of Japan Has Developed the First RFID Chip That Can Operate at All Frequencies from 13.56 MHz to 2.45 GHz," (http://www.rfidjournal.com/article/articleview/831/).

[5] **Altria Group, Inc.**, *Cooperation Agreement between PMI and the European Commission* (2005) (http://www.altria.com/AnnualReport/ar2005/2005ar_07_0207.aspx).

[6] **Altria Group, Inc.**, PMI Annual Report (2005) (http://www.altria.com/Annual Report/ar2005/2005ar_07_0207.aspx).

[7] **Jacoby, D.**, "RFID Comes of Age," Executive Briefing, Economist Intelligence Unit in partnership with Harvard Business School Publications (Economist Intelligence Unit, London, 2006).

[8] **Womack, J. P.**, **Jones, D. T.**, and **Roos, D.**, *The Machine That Changed the World: The Story of Lean Production* (Harper Perennial, New York, 1991).

[9] **Chan, R.**, **Ram, S.**, and **Miles, S.**, "Data Exchange WG Pilot Learnings," in *EPCglobal Cookbook* (http://www.epcglobalinc.org/what/cookbook/; http://www.epcglobalna.org/SubscriberTools/InternalImplementationPlanning/EPCRFIDCookbook/tabid/220/Default.aspx).

[10] **Heinrich, C. E.**, *RFID and Beyond: Growing Your Business through Real World Awareness* (Wiley, New York, 2005).

Appendix – links to RFID technology and applications resources

I.   RFID interface specifications

- **ISO 18000-7 and ISO 18000-6c HF and UHF RFID standards** (Ch. 1):
  http://www.iso.org/iso/en/CatalogueDetailPage.CatalogueDetail?CSNUMBER=34117.
- **EPCglobal Class 1 Generation 2 UHF air interface protocol specifications** (Chs. 1 and 2):
  http://www.epcglobalinc.org/standards/uhfc1g2.
- **EPCglobal EPCIS specifications and open-source implementations** (Chs. 1 and 15):
  http://www.epcglobalinc.org/standards/.

  Accada – an open-source Java EPCIS implementation and RFID reader simulator from the Auto-ID Labs at MIT and the ETH Zürich (http://www.accada.org/).

  Mentor (MIT EPC Net) – an open-source .NET EPCIS from the Auto-ID Labs at MIT; reference code for EPCIS compatibility (http://source.mit.edu/).

- **A proposal for incorporating RFID plus sensor data** (Ch. 5):
  IEEE 1451.7 integration with ISO 24753 (http://ieee1451.nist.gov/http://www.iso.org/iso/en/CatalogueDetailPage.CatalogueDetail?CSNUMBER=38840&scopelist=PROGRAMME).

II.   Test capabilities (in the order of chapters as presented)

- **Semiconductor design and prototyping** (Ch. 3):
  Hao Min, Director, Fudan University Auto-ID Labs (http://www.autoidlab.fudan.edu.cn/).
- **RTLS design and test** (Ch. 6):
  Kaveh Pahlavan, Director, Worcester Polytechnic Center for Wireless Information Network Services (CWINS) (http://www.cwins.wpi.edu/).
- **Medical device RF test center** (Ch. 8):
  Gisele Bennett, Director, Georgia Tech Logistics and Maintenance Applied Research Center (http://landmarc.gtri.gatech.edu/).

- **EPCglobal certified test center** (Ch. 9):
  Bill Hardgrave, Executive Director, University of Arkansas Information Technology Research Institute, RFID Research Center (http://itri.uark.edu/rfid/).
- **RFID cold chain distribution center** (Ch. 11):
  J. P. Emond, Co-Director, University of Florida Center for Food Distribution and Retailing (http://cfdr.ifas.ufl.edu/).
- **Lab for computer-aided design and production (LICP)** (Ch. 13):
  Dimitris Kiritsis, Associate Director, EPFL, Lausanne (http://licpwww.epfl.ch/main.asp).

III.   Simulation environments

- **Simulation analyses for UHF and HF tag performance** (Ch. 4):
  Marlin Mickle, Director, The University of Pittsburgh RFID Center of Excellence (http://www.engr.pitt.edu/SITE/RFID/).
- **EPC network simulator**; sponsored research project with SAP Research:
  John R. Williams, Director, Massachusetts Institute of Technology Auto-ID Labs (http://autoid.mit.edu/).
- **RFID tag selection tool** balancing timeliness or read rate against accuracy and completeness (Ch. 10):
  Duncan McFarlane, Director, Cambridge University Auto-ID Labs (http://www.autoidlabs.org.uk/).
- **Attack models for anti-counterfeiting** (Ch. 12):
  Elgar Fleisch, Director, the Auto-ID Labs at the University of St. Gallen and the ETH Zürich (http://vsgr.inf.ethz.ch/autoidlabs.ch/).
- **RFID application and demonstration Center** (Ch. 14):
  Bernd Scholz-Reiter, Ph.D., Director, University of Bremen, Division of Intelligent Production and Logistics Systems (IPS), Bremen Institute of Industrial Technology and Applied Work Science (BIBA) (http://www.biba.uni-bremen.de/ips.html?&L=1).

Editor biographies

Stephen B. Miles

In 2004 Steve Miles joined the Auto-ID Labs at MIT, where he currently serves as RFID Evangelist and Research Engineer. Steve has over 15 years of computer systems integration and network services experience. His research interests are in identifying industry requirements for electronic product code (EPC) data exchange. Steve is a 2003 graduate of the Management of Technology executive MBA program at MIT Sloan, during which time he also served as a consultant in value-added IP services strategy to leading service providers in the USA and Europe.

Previously Steve served on the executive teams of Wireless IP Networks, a wireless softswitch start-up, IronBridge Networks, the terabit MPLS router OEM, NMS Communications (NMSS), a supplier of VoIP and SS7 PC-based telephony platforms to wire line and wireless OEMs, and Officenet Inc., where he founded a regional third-party computer services leader, which was acquired by Decision One in 1996.

Steve participates in EPCglobal, IEEE, W3C, and healthcare industry standards organizations, is a frequent speaker at RFID conferences and chairs the MIT Enterprise Forum RFID SIG. As the founder and Co-chair of the RFID Academic Convocation, Steve is working to foster collaboration with industry, government, and academic researchers to address RFID issues requiring cross-organizational and cross-continent collaboration.

Professor Sanjay Sarma

Sanjay Sarma is an associate professor at MIT. Sarma was one of the founders of the Auto-ID Center at MIT, which developed many of the technical concepts and standards of retail. He also chaired the Auto-ID Research Council grouping six labs worldwide, which he helped set up. Today, the suite of standards developed by the Auto-ID Center, commonly referred to as the EPC, is being used by over a thousand companies on five continents. Sarma serves on the board of EPCglobal, the worldwide standards body he helped create. Between 2004 and 2006, Sarma took a leave of absence from MIT to found a software company called OAT Systems. He continues to serve on the board of OAT.

Sarma received his B.Sc. from the Indian Institute of Technology, his M.Sc. from Carnegie Mellon University, and his Ph.D. from the University of California at Berkeley. In between degrees, Sarma worked at Schlumberger Oilfield Services in Aberdeen. Sarma's M.Sc. thesis was in the area of operations research, and his Ph.D. was in the area of manufacturing automation. His current research projects are in the areas of radio frequency identification, IC packaging, manufacturing, CAD/CAM, machine design, RFID applications, device networking, and smart devices. He has over 50 publications in computational geometry, virtual reality, manufacturing, CAD, RFID, security, and embedded computing.

Sarma is a recipient of the National Science Foundation CAREER Award, the Cecil and Ida Green Career Development Chair at MIT, the Den Hartog Award for Excellence in Teaching, the Keenan Award for innovations in undergraduate education, the New England Business and Technology Award, and the MIT Global Indus Award. He was selected on 2003's *Business Week* ebiz 25 and *Fast Company Magazine*'s Fast Fifty.

Professor John Williams

John Williams is Professor of Information Engineering in Civil and Environmental Engineering and specializes in large-scale simulation and computation. He is Director of the MIT Auto-ID Laboratory, which is one of seven university laboratories worldwide using RFID to design the architecture of "The Internet of Things." The challenges involve building a global identity system for entities that is secure and scalable.

Williams was previously Director of MIT's System Design and Management Program and has lectured widely on online education. He has also worked in industry and was Vice President of Engineering at two software start-ups. He is currently doing research on RFID networks for SAP, Intel, EPCglobal, and Microsoft. He teaches two graduate courses on Web System Architecting and on Modern Software Development.

Williams holds an M.A. in Physics from Oxford University, an M.S. in Physics from UCLA, and a Ph.D. in numerical methods from the University of Wales, Swansea and has published two books and over 100 journal and conference papers. In his youth he was an Oxford rugby Blue and played for Bristol and Newport. He now plays golf and dances Argentine tango. He was voted one of the 50 most powerful people in computer networking in 2005.

Index

13.56 MHz, 1, 2, 48, 49, 55, 149, 151, 187
2.4–2.48 GHz, 105
433 MHz, 13, 62, 110, 187, 189, 195, 206
915 MHz, 49, 50, 55, 149, 153
access management, 163, 164, 169
actuators, xx, 61, 63, 65, 67, 68, 72, 176, 190
address services, 163
aerospace, xvi, xvii, xix, 88, 122, 123, 127, 129, 130, 132, 133, 134, 136, 140, 141, 143, 144, 145, 146, 148, 149, 151, 157, 198, 200, 205
air interface, 55, 56, 62, 63, 103, 106, 195
anti-counterfeiting, xvii, 87, 159, 160, 161, 162, 163, 165, 167, 204, 207, 208
AOA, 74, 75, 76, 77, 83, *see* Location Techniques
attack model, 157, 159, 160, 162, 167
autonomic logistics, xvii, 199, 209
available energy, 159

baggage, xvii, 3, 30, 54, 148
blood, 11, 30, 69, 148, 149, 152, 154
Bluetooth, xviii, 4, 67, 187, 204, *see* IEEE 802.15.1
branded sportswear, 157
bulk reading, 162, 186
business processes, xvi, xvii, xviii, 12, 13, 28, 54, 119, 131, 134, 148, 174, 209, 210

carton-level, 165
challenge–response protocols, 157
chemical markers, xvii, 162
cold chain, xvii, xix, 30, 68, 105, 107, 144, 145, 146, 147, 148, 154, 155, 156, 157, 200
communication bandwidth, 159, 165
computational power, 159, 160
conductive polymers, 55
control systems, 3, 9
counterfeiting, 27, 89, 116, 157, 158, 164, 204, 208, 211, 215

data models, xvii, 198
data rate, 39, 55, 56, 63
data standards, 162
denial-of-service attacks, 162
designer clothing, 157

digital media, 158

electromagnetic, 3, 4, 6, 9, 20, 38, 53, 54, 55, 105, 107, 108, 110, 117, 198, 200
EPC network, xvi, xix, 9, 202, 203
EPCIS, 2, 8, 9, 10, 13, 19, 21, 23, 32, 101, 202, 203, 204, 213

fashion accessories, 158
Finite Element Method, 54
FMCG, 48, 59, 60
frequency spectrum, 62, 105, 159
fresh produce, xvii, 146, 151, 154

gap analysis, 102, 199
garments, 54
ghost tags, 2
GPS, 74, 75, 84, 186, 189, 190, 206, 210

hazardous waste, xvii, 177, 209
healthcare, xii, xix, xxi, 1, 9, 199, 203, 210, 212
holograms, 157, 161, 162

IEEE 1451, xvi, 62, 63, 64, 65, 66, 67, 68, 69, 70, 71, 72, 74, 204, 205, 206, 213
IEEE 802.11, 4, 7, 75, 77, 204
IEEE 802.11g, xvi
IEEE 802.15.4, 204, *see* Zigbee
impedance, 50, 54, 55, 64, 66, 67
intangible assets, 157
Intellectual Property Rights, 158
interoperability, xv, xv, xviii, xviii, xix, 3, 8, 47, 56, 63, 67, 75, 105, 107, 176, 203
interrogators, xviii, 5, 6, 68, 201, 208
ISO 18000-7, xviii, 62, 187
ISO 24730, xvi, 206, 211
item level, 54, 59, 113, 116, 117, 118, 119, 121, 163, 195, 201, 208, 211

level of security, 159, 165
localization techniques, 75, *see also* AOA, RSS, TOA, TDOA

location tracking, xvi, 9, 74, 75, 101, 104, 189,
  see RTLS

MAC, 75
maximum gate count, 159
Media Access Control, 56
medical, xvi, xix, 61, 63, 107, 110, 111, 113, 145,
  199
MEM's, xviii, 4
memory size, 159, 187, 188, 201
micro printings, 157, 161
mixed tote, 60
multipath, 2, 5, 74, 75, 76, 77, 79

network layering, xix, 55
NFC, 161, 208

Obfuscation, 164
Object Specific Security, 54

pallet-level, 149
perishables, 157, 211
pharma, xvii, 59, 61, 155
pharmaceuticals, 48, 88, 145, 157, 158, 202
phy, 65, *See* Physical Layer
physical layer, 55, 57, 63, 72, 202
plausibility checks, 163
plug-and-play, 61, 64, 66
point of sale, 60, 117, 137, 171
product-specific data, 164
propagation, 3, 34, 50, 74, 75, 80, 83, 84, 87, 105,
  106
protocol layer, 58
public-key crypto systems, 160

radio direction-finding, 74
RDF, xvii, 173, 179, 180, 181, 185, 213
read ranges, 104, 105, 160, 162, 165
read rates, 31, 33, 59, 68, 108, 114, 115, 162, 201
removal–reapplication attacks, 161
resistance, 54, 164
retail, xvi, xix, xx, 4, 5, 12, 27, 28, 29, 33, 47, 87,
  95, 113, 114, 116, 118, 120, 129, 136, 145, 146,
  147, 148, 154, 161, 198, 207, 210

return on investment, 59, 114, 135, 158
risk, 107, 110, 158, 166
RSS, 74, 75, 76, 77, 78, 80, 83
RTLS, xvi, xviii, 75, 80, 84, 101, 108, 198, 199,
  204, 206, 210, 214

search, xvii, 3, 4, 5, 12, 56, 57
secure authentication, 165
security features, 157, 159, 161, 162
security models, 159
security/privacy, 60
Semantic Web, xvii
sensor networking, xvi, xviii, 71
shadow fading, 74, 78
signature value, 164, 164, 165
simulation methodology, xvi
skin effect, 54
spare parts, 149, 158, 162

tag cloning, 161
tag ID, 71, 163
tag omission, 161
tax income, 158
TDOA, 74, 75, 76, 77, 189
telemetry, 8, 210, 211
temperature sensors, xvii, 29, 33, 47, 68, 69, 200
temperature tracking, 30, 68, 149, 154
TFTCs, xviii, 4
TOA, 74, 75, 76, 77, 80, 83
tobacco products, 158
traceability, 17, 120, 144
track and trace, 91, 135, 149, 161, 163, 164
transducer, 9, 64, 65, 66, 67
transponders, xviii, 2, 3, 33, 54, 75, 159, 160, 162,
  163, 165, 166, 186, 187, 189, 192, 200, 204

validated shipping containers, 154
validation key, 164, 165
validation process, 147, 154
vials, 54

WPAN, xix, 75, 205

Zigbee, xviii, 4